W9-AFO-011

# Labeling Genetically Modified Food

# Labeling Genetically Modified Food

## The Philosophical and Legal Debate

EDITED BY

Paul Weirich

2007

OXFORD
UNIVERSITY PRESS

Oxford University Press, Inc., publishes works that further
Oxford University's objective of excellence
in research, scholarship, and education.

Oxford New York

Auckland Cape Town Dar es Salaam Hong Kong Karachi
Kuala Lumpur Madrid Melbourne Mexico City Nairobi
New Delhi Shanghai Taipei Toronto

With offices in

Argentina Austria Brazil Chile Czech Republic France Greece
Guatemala Hungary Italy Japan Poland Portugal Singapore
South Korea Switzerland Thailand Turkey Ukraine Vietnam



Published by Oxford University Press, Inc.
198 Madison Avenue, New York, New York 10016

www.oup.com



Library of Congress Cataloging-in-Publication Data
Labeling genetically modified food : the philosophical and legal debate / edited by Paul Weirich.
p. cm.
Includes bibliographical references and index.
ISBN 978-0-19-532686-4
1. Genetically modified foods.
2. Food—Labeling.
3. Consumer's preferences. I. Weirich, Paul, 1946–
TP248.65.F66L33 2007
363.19'29—dc22 2007001931

9 8 7 6 5 4 3 2 1

Printed in the United States of America
on acid-free paper

To Michèle

# Preface

Agrobiotechnology holds great promise for food production, but consumers want to be apprised of its introduction into the supermarket. Policies on labeling genetically modified (GM) food vary, and the differences create friction. The United States does not require labeling, and the European Union does. Which labeling policy is better?

This collection thoroughly surveys the case for and against labeling GM food. The authors—with perspectives from philosophy, law, agricultural economics, and the life sciences—have diverse opinions about labeling policy and the legal principles that underlie it. Their essays create a springboard for new ideas concerning consumer information.

The collection is the first book-length treatment of the issue of labeling GM food. The essays are brand new except for two, which have been updated to cover the latest developments in labeling policy. Ample references direct the reader to additional literature. They are a valuable resource for students and experts in the variety of disciplines covering regulatory policy.

A conference at the University of Missouri–Columbia, held November 4–5, 2005, generated this collection. Missouri is an agricultural state, home of the agbiotechnology firm Monsanto. The University of Missouri–Columbia is a leader in genetic research on soybeans, sorghum, and swine. The university organized the conference on food labeling as a service to the state, which benefits from an objective, impartial discussion of labeling. The conference had these university sponsors: the College of Agriculture, Food, and Natural Resources; the Office of Research; the Kline Missouri Chair in the Department of Philosophy; the Life Sciences and Society Program; the College of Arts and Science; the Office of the Provost; the Department of Philosophy; and the Life Sciences Center. Special thanks go to Eric Roark for assisting with conference arrangements and to the international group of faculty, graduate students, consumer advocates, and professionals in government and biotechnology who expressed interest in the conference and its proceedings.

The Research Council at the University of Missouri–Columbia supported preparation of this book. Ted Poston and Kenneth Boyce expertly assembled the essays to form a unified collection. Christopher Haugen prepared the index. Suggestions for the jacket artwork came from Blake Dinsdale, Eliot Hermann, and the National Center for Soybean Biotechnology at the University of Missouri–Columbia. Brian Desmond shepherded the manuscript through the production process. The project would not have succeeded without the encouragement of Oxford University Press's humanities editor, Peter Ohlin.

# Contents

# Contributors

**Carl Cranor** is professor of philosophy at the University of California–Riverside, specializing in legal and moral philosophy. He has written widely on philosophic issues at the intersection of science and the law. Some of that research focused on philosophic issues in risk assessment and the regulation of toxic substances. Some has been concerned with philosophical analysis of the acceptability of risks as well as with issues concerning genetically modified organisms. More recent research concerns philosophic issues on the use of science in the tort law as well as work clarifying aspects of the Precautionary Principle. He has written *Regulating Toxic Substances: A Philosophy of Science and the Law* (Oxford University Press, 1993), as well as numerous articles in philosophy, science, and law journals. He has just completed *Toxic Torts: Science, Law and the Possibility of Justice* (2006). As a Congressional Fellow, he worked at the U.S. Congress's Office of Technology Assessment, co-authoring *The Identification and Regulation of Carcinogens* (1987). He also served on California's Proposition 65 Science Advisory Panel as well as its Electric and Magnetic Fields Science Advisory Panel. He is an elected fellow of the American Association for the Advancement of Science and the Collegium Ramazzini. He received his B.A. degree from the University of Colorado (mathematics), a Ph.D. from UCLA (philosophy), and an M.S.L. (Master of Studies in Law) from Yale Law School.

**Fred H. Degnan** is a partner in King & Spalding's Washington, D.C. office, where since 1988 he has specialized in food and drug law. Prior to coming to King & Spalding, he served for 11 years in the FDA's Office of Chief Counsel, where he had substantial experience with FDA litigation and enforcement and served as the agency's associate chief counsel for foods. He received the agency's highest awards. At King & Spalding, he represents several large multinational food companies, a number of domestic drug producers, several international science-based non-profit associations, and several biotechnology companies. Since 1989, in addition to his responsibilities at King & Spalding, he has taught food and drug law at the Catholic University of America's Columbus School of Law, where he serves as a Distinguished Lecturer.

**Margaret (Peggy) Rosso Grossman** is Bock Chair and professor of agricultural law in the Department of Agricultural and Consumer Economics, University of Illinois at Urbana-Champaign. She teaches courses in agricultural law, environmental law, and veterinary jurisprudence. She is author of numerous U.S. and European law review articles and book chapters on agricultural and environmental law topics. She is co-author or co-editor of several books published in Europe, most recently *Agriculture and International Trade: Law, Policy and the WTO* (2003). As a member of the National Research Council Committee on Air Emissions from Animal Feeding Operations, she contributed to two research reports (2002, 2003). She has received three Fulbright Senior Scholar Awards to support her research in Europe: Western Europe (1986–1987), the Netherlands (1993–1994), and European Union Affairs (2000–2001). She was a German Marshall Fund Research Fellow (1993–1994), served as president of the American Agricultural Law Association (1991), and received the 1993 AALA Distinguished Service Award (1993). She was elected to the European Council for Agricultural Law and received the Silver Medal from that organization in 1999 for her contributions. She is a frequent visiting research professor in the Law and Governance (formerly Agrarian Law) Group at Wageningen University, the Netherlands. She is a graduate of the University of Illinois (B.Mus., 1969; Ph.D. [musicology], 1977; J.D. summa cum laude, 1979) and Stanford University (A.M. [musicology], 1970).

**Nicholas Kalaitzandonakes** is the MSMC Endowed Professor of Agribusiness in the Department of Agricultural Economics and director of the Economics and Management of Agrobiotechnology Center at the University of Missouri–Columbia. His research, teaching, and outreach focus on the economics and policy of agrifood biotechnology and other agribusiness innovations.

**Thomas A. Lambert** is an associate professor at the University of Missouri–Columbia School of Law. He joined the law faculty in the fall of 2003 from the Chicago law firm of Sidley Austin Brown & Wood, where he practiced antitrust litigation from 2000 to 2003. Prior to entering law school, he was an environmental policy analyst at the Center for the Study of American Business at Washington University in St. Louis. He then attended the University of Chicago Law School, where he was a Bradley Fellow and served as comment editor of the *University of Chicago Law Review*. After graduating with honors in 1998, he clerked for Judge Jerry E. Smith of the U.S. Court of Appeals for the Fifth Circuit. He then spent a year as the John M. Olin Fellow at Northwestern University Law School. Lambert's published work has appeared in scholarly journals, such as *Georgia Law Review*, *Yale Journal on Regulation*, and *Journal of Labor Research*, and in more popular policy journals, such as *Public Interest* and *USA Today Magazine*. He teaches environmental law, antitrust law, and business organizations.

**Thomas O. McGarity** holds the W. James Kronzer Chair in Trial and Appellate Advocacy at the University of Texas School of Law. He has taught environmental law, administrative law, and torts at UT Law School since 1980. Prior to that, he taught at the University of Kansas School of Law. After clerking for Judge William E. Doyle of the Federal Court of Appeals for the Tenth Circuit in Denver, Colorado, McGarity served as an attorney-advisor in the Office of General Counsel of the Environmental Protection Agency in Washington, D.C. He has written widely in the areas of environmental law and administrative law. He has written two books on federal regulation: *Reinventing Rationality* (1991) describes and critiques the implementation of regulatory analysis and regulatory review requirements that were put into place during the Carter and Reagan administrations. *Workers at Risk* (1993; co-authored with Sidney Shapiro) describes and critiques the implementation of the Occupational Safety and Health Act during its first 20 years. McGarity is president of the Center for Progressive Reform, a nonprofit organization consisting of scholars who are committed to developing and sharing knowledge and information, with the ultimate aim of preserving the fundamental value of the life and health of human beings and the natural environment.

**Peter Markie** is a professor of philosophy at the University of Missouri–Columbia. The author of *Descartes's Gambit* (1985) and *A Professor's Duties* (Rowman and Littlefield, 1994), he has published papers in the history of philosophy, ethics, the philosophy of mind, and epistemology.

**Leonie A. Marks** is research assistant professor in the Department of Agricultural Economics, director of the University of Missouri-Columbia Life Sciences and Society Program, and program director at the Economics and Management of Agrobiotechnology Center at the University of Missouri–Columbia. Her research and teaching focus on the economics and policy of agrifood innovation and consumer behavior.

**Michael W. Pariza** is director of the Food Research Institute and Wisconsin Distinguished Professor of Food Microbiology and Toxicology, University of Wisconsin–Madison. He has authored or co-authored more than 150 articles and publications, holds more than 25 U.S. patents, and is recognized by Thompson-ISI as one of the most highly cited researchers of the last two decades. He received his B.S. in bacteriology at the University of Wisconsin–Madison and his M.S. and Ph.D. in microbiology at Kansas State University. He completed three years of postdoctoral study at the McArdle Laboratory for Cancer Research at the University of Wisconsin–Madison and joined the faculty of the Food Research Institute in 1976. The research program of the Food Research Institute centers on food-borne microbial pathogens and toxins; Pariza's research focuses on conjugated linoleic acid, the biological activity of which was discovered in his laboratory, and on various food toxicology issues, most recently the control of acrylamide formation in food.

**Philip G. Peters** is the Ruth L. Hulston Professor of Law at the University of Missouri–Columbia School of Law, where he specializes in health law. He was also the founding director of the University of Missouri-Columbia Biotechnology and Society Program, an interdisciplinary initiative assembled to study the social and legal implications of modern genetic technology. He was recently a visiting fellow at Clare Hall, Cambridge University, where he completed a book laying out a framework for the regulation of risky reproductive technologies—*How Safe Is Safe Enough?* (Oxford University Press, 2004). In 2004–2005, he was the reporter for a committee established by the National Conference of Commissioners on Uniform State Laws to study the need for uniform state laws to prevent the misuse of genetic information. He also writes about malpractice reform and the law of death and dying.

**R. Michael Roberts** is a Curators' Professor at the University of Missouri–Columbia, with appointments in animal sciences, biochemistry and veterinary pathobiology. He is currently an investigator in the University of Missouri Christopher S. Bond Life Sciences Center. He gained his B.A. and D.Phil. in Plant Sciences from Oxford University but since the mid-1970s has worked primarily as a reproductive biologist. His interests are in early pregnancy in mammals, particularly how the early embryo signals its presence to the mother through the production of small proteins called interferons. His work is supported primarily through federal agencies such as the National Institutes of Health and the U.S. Department of Agriculture. He has published more than 250 papers in refereed scientific journals and 50 reviews and chapters in books. He was elected to the National Academy of Sciences in 1996 and has received several international awards, including the Milstein Prize for research on interferons and the Wolf Prize for agriculture. Roberts was chief scientist with the USDA's Competitive Grants Program (the National Research Initiative) from 1998 to 2000, a period in which the controversies surrounding genetically modified organisms were at their height. He also served on the National Research Council's committee that published recommendations to the FDA on concerns regarding the use of genetically modified animals for food (*Animal Biotechnology: Science Based Concerns*, 2002).

**Alan Rubel** is currently the law clerk to Justice Ann Wash Bradley of the Wisconsin Supreme Court. He received his Ph.D. in philosophy at the University of Wisconsin–Madison in 2006 and his J.D. (magna cum laude) from the University of Wisconsin Law School in 2003. His recent publications include "Protecting the Autonomy of European and American Consumers: The Case for Labeling on GE Foods" (*The Journal of Agricultural and Environmental Ethics*, 2005) and "Privacy and the USA Patriot Act: Rights, the Value of Rights, and Autonomy" (*Law and Philosophy*, 2007).

**Robert Streiffer** is an associate professor at the University of Wisconsin–Madison. He has appointments in the Department of Medical History and Bioethics in the Medical School and the Department of Philosophy in the College of Letters and Sciences, and is affiliated with the Department of Medical Sciences in the School of Veterinary Medicine, and the Department of Agricultural and Applied Economics in the College of Agricultural and Life Sciences. He received his Ph.D. in ethics from MIT in 1999, and his research now focuses on ethical and policy issues arising from modern biotechnology, both agricultural and biomedical. Recent publications include the book *Moral Relativism and Reasons for Action* (2003); "Academic Freedom and Industry-Imposed Restrictions on Academic Biotechnology Research" (*The Kennedy Institute of Ethics Journal*, 2006), "At the Edge of Humanity: Human Stem Cells, Chimeras, and Moral Status" (*The Kennedy Institute of Ethics Journal*, 2005), "An Ethical Analysis of Ojibway Objections to Genomics and Genetics Research on

Wild Rice" (*Philosophy in the Contemporary World*, 2005), and "The Political Import of Intrinsic Objections to Genetically Engineered Food" (*The Journal of Agricultural and Environmental Ethics*, 2005).

**Steven S. Vickner** is associate professor in the Department of Economics at Utah State University. His research and teaching interests focus on the economics of the global food industry.

**Paul Weirich** is a professor in the Philosophy Department at the University of Missouri–Columbia. He earned a Ph.D. at UCLA and held a prior teaching position at the University of Rochester. He is a past chair of his department and teaches primarily logic and decision theory. As a graduate student, his interest in probability prompted a dissertation treating subjective probability in decision theory. He has written three books on related topics: *Equilibrium and Rationality: Game Theory Revised by Decision Rules* (1998), *Decision Space: Multidimensional Utility Analysis* (2001), and *Realistic Decision Theory: Rules for Nonideal Agents in Nonideal Circumstances* (Oxford University Press, 2004). Currently, he is an associate editor for *The New Dictionary of Scientific Biography*, an eight-volume collection to be published by Scribner, and edits its entries on decision and game theory.

**Clark Wolf** is Director of Bioethics at Iowa State University. Before joining ISU he spent 10 years as a member of the philosophy department at the University of Georgia. He holds degrees in musicology and music performance from Oberlin Conservatory of Music and pursued graduate work in economics at the University of Pittsburgh and in philosophy at the University of Arizona. Much of his published work is in political and legal philosophy and in the philosophy of economics. His works in progress include papers on ethics and biotechnology, intellectual property, intergenerational justice and global environmental change, and the market for human transplant organs.

# Introduction

Consumers are reasonably interested in many facts about food products, from safety to the effects of production on the environment and the economy. Informing consumers is an important social goal, and finding the appropriate means of meeting this goal is a pressing, complex social issue. For example, making a food label informative but not misleading is a complicated matter. Given that most consumers believe that food products do not contain genetically modified (GM) ingredients, is a food label misleading if it does not correct that impression? If a food label states that a food product does not contain GM ingredients, does the label misleadingly suggest that such ingredients are unsafe? A satisfactory position on labeling GM food must attend to these and a host of other complications.

This volume presents the basic science of GM food and then a spectrum of philosophical and legal views about labeling. It guides the reader to an informed opinion about labeling.

Genetically modified corn and soybeans grow on many farms. Some consumers fear the consequences of eating GM food, also known as genetically engineered or bioengineered food. Others fear that GM food crops will damage the environment and hope to vote against them at the supermarket. Various groups want the U.S. Food and Drug Administration (FDA) to require food labels to indicate GM ingredients. The European Union requires such labeling, but the FDA has resisted the arguments for it. The FDA contends that its charter instructs it to label only factors that are significant for health, nutrition, or use. Scientific studies detect no substantial difference between food from traditional crops and from GM crops. The FDA regards genetic modification as immaterial. Hence, in its view, it does not have authority to demand that food labels attend to it. So far, it has issued guidelines only for voluntary labeling of food with GM ingredients.

Corporations and farmers investing in GM crops point out the benefits. Genetically modified corn is resistant to insect pests. Genetically modified soybeans are resistant to herbicides used to kill weeds. If food labels announce GM ingredients, consumers may shun products with those ingredients. If consumers turn away from those products, the industry introducing genetic modifications will suffer. The benefits of GM crops may be postponed or lost altogether.

The issue brings to the fore problems in democratic theory. Some forms of democracy rely on elections of representatives. The representatives become informed about proposed laws and use their expertise to promote the public's interests. Other forms of democracy rely on referenda and additional methods of direct decision making by the public. Advocates of representative democracy claim that it is more efficient to let experts decide on behalf of the public than to require each citizen to become informed before casting a vote in a referendum. Advocates of participatory democracy hold that experts are divided on complex issues and that relying on them invites manipulation by special interest groups and lobbyists. Should the FDA respond to the public's wishes, or should it proceed along the path that in its view best serves the nation's interests? Such dilemmas are challenging for democratic governments.

Genetically modified food also raises questions about the objectives of food labeling policy. Tuna may be labeled to indicate fishing techniques. Labels may say whether fishing boats used nets that allowed dolphins to escape. These matters do not affect health or nutrition. Should the reasoning behind such labeling license consumers' use of labeling to promote other environmental causes? Labeling food for GM ingredients gives environmentalists leverage. They may avoid foods with those ingredients not because of worries concerning health or nutrition, but because of worries that GM crops will alter traditional crops and wild plants through pollen drift. Does the type of reason that warrants voluntary labeling of dolphin-safe tuna also warrant voluntary labeling of GM food? May it, in addition, justify mandatory labeling? The step from voluntary to mandatory labeling is large.

Labels provide information about food more effectively than do websites, the media, and advertisements. However, labeling may be costly. Requiring labeling for GM ingredients requires keeping track of GM corn in grain elevators and throughout food production. The expense may raise the cost of food significantly. To keep down costs, a label might say only that a product *may* contain GM ingredients. However, such vagueness will not satisfy concerned consumers who want to know which products contain GM ingredients. Providing that knowledge is feasible but not simple.

Some essays in this collection support current labeling policies, some urge improvements in voluntary labeling, and some back mandatory labeling. The essays consider moral, institutional, and economic reasons for and against labeling. The reasons concern health, the environment, consumer autonomy, labeling costs, trade, and global food production. Many considerations bear on the provision of information to consumers.

In chapter 1, Michael Pariza considers various definitions of food and associated approaches to labeling it. A biological perspective explains the origin of dilemmas of food production, such as selection of methods of protecting from pests crops bred for their nutritional value. To illustrate the volatility of debate about GM food, Pariza reviews the introduction and removal of GM tomato products in the United Kingdom during the 1990s. Society's interests require impartial assessment of scientific evidence and philosophical and legal arguments bearing on labeling policy.

Michael Roberts observes in chapter 2 that genetic modification of food crops is routine and that no scientific obstacle blocks genetic modification of animals used for food. Before GM foods reach the market, regulators demand tests of their safety and the environmental impact of their production. The tests of their safety examine their chemical composition. Labeling GM ingredients presumes tests of their presence, and such tests are available, although highly sensitive tests are costly. Whether to label GM food products is a political issue assisted but not settled by scientific considerations.

Fred Degnan describes the origin of the FDA's current labeling policy and its main features in chapter 3. The FDA prohibits putting some claims on food labels. Labels may not be false or misleading. Furthermore, the FDA requires putting some information on food labels. Labels must include the name of the food, the name and address of the manufacturer, a measure of the food's quantity, its ingredients, and its nutritional value. Hence, if the fatty acids in soy oil from GM plants differ in composition from the fatty acids in conventional soy oil, then the GM oil must carry a label indicating the difference. Degnan holds that the FDA's understanding of its legal authority is correct. Unless there is some material difference between a GM food and its traditional counterpart, the agency does not have the authority to require that the food's label say that it is genetically modified. If such a material difference exists, he believes that the agency may and should require labeling. He also believes that nonmisleading voluntary labeling is a lawful way to inform consumers of nonmaterial but nevertheless desired information about food.

In chapter 4, Margaret Rosso Grossman describes labeling in the European Union. The European Union has approved some GM food and feeds. Nonetheless, most consumers do not buy GM food, and some regions ban it. Cultivation of GM crops is limited. The European Union relies on the Precautionary Principle, which urges caution, and so the European Union has many regulations to reduce risks to human health, to animal health, and to the environment. Regulations distinguish food from GM plants, such as oil from GM soybeans, and food produced using GM plants, such as meat from cattle fed GM corn. The European Union insists on labeling food from GM plants if it constitutes more than a small percentage of a food product. Farmers are preparing for the coexistence of GM and non-GM crops. In Denmark, they pay into a fund that will compensate for contamination of non-GM crops by GM crops. The European Union's regulations on GM food and feed impose costs on U.S. farmers exporting to the European Union. They must separate GM and traditional crops. Countries around the globe may facilitate trade by adopting uniform regulations.

Robert Streiffer and Alan Rubel argue in chapter 5 for mandatory labeling of GM food. They discuss, in particular, food from GM animals, such as GM salmon and GM pigs, which may reach the market soon. Some consumers may reject food products from GM animals because of a concern for animal welfare. Streiffer and Rubel examine the principles underlying labeling regulations. They conclude that labeling policy should protect not just health and safety but also consumer autonomy. They hold that agencies such as the FDA have a democratic obligation to require labeling to protect consumer autonomy. Most consumers do not realize that many food products have GM ingredients, and Streiffer and Rubel believe that it is misleading not to label those products. Labeling permits consumers to exercise choice about issues that matter to them. Streiffer and Rubel summarize data showing that consumers are resistant to GM food and want it labeled.

In chapter 6, Peter Markie criticizes the argument from consumer autonomy to mandatory labeling, which concludes that manufacturers have an obligation to inform consumers about GM ingredients and that legislators should require fulfillment of that obligation. The obligation to label is, at most, a prima facie obligation. Other factors may override it. Is labeling the best means of promoting consumer autonomy? The Internet provides more information about a product than a label can. Manufacturers need not put on a label all the information consumers want. Required information should be limited to information bearing on health or other social priorities. In typical cases, manufacturers have no overall moral obligation to provide information about GM ingredients. They legitimately profit from the sale of GM food products as long as their marketing is not deceptive.

In chapter 7, Nicholas Kalaitzandonakes, Leonie Marks, and Steven Vickner present market data showing that consumers are not willing to pay a premium for non-GM food when given a choice between GM and non-GM food. Their data come from the Netherlands and China, where GM and non-GM foods are labeled. In the Netherlands, consumers do not respond to labels on processed food with information about GM ingredients. In China, consumers do not respond to labels on soy oil that say it comes from GM soybeans. The authors believe that preferences consumers reveal in the marketplace are more significant than preferences consumers state in opinion surveys. They conclude that mandatory labeling does not bring benefits that justify its cost. Voluntary labeling can accommodate the minority who care about the difference between GM and non-GM foods.

Thomas McGarity advocates labeling in chapter 8. Despite efforts to prevent the presence of allergens in GM food, that food may contain allergens. Shouldn't a risk-averse consumer be given the option of avoiding GM food? McGarity observes that consistency in labeling policy creates a case for labeling GM food. Governments commonly create regulatory agencies to help consumers acquire information. The FDA has authority to require labeling GM food because a food's being genetically modified is material in the relevant sense. Materiality may concern the process of production as well as nutrition, health, and use. The constitutionally protected right of free speech, contrary to the opinion of some courts, does not give a corporation a right to be silent about GM ingredients.

Philip Peters and Thomas Lambert argue in chapter 9 for voluntary labeling of GM food but against mandatory labeling. The National Academy of Sciences recognizes that GM food is riskier than conventional food because of long-term risks. Some rational consumers may want to avoid GM food. However, if informed consumers buy GM food, then they accept the risks in exchange for the benefits. Voluntary labeling accommodates consumer aversion to risk but is less restrictive than mandatory labeling. In a socially efficient way, it allows market forces to settle how much information about GM food manufacturers offer consumers. Voluntary labeling works for organic and kosher food and can work for GM food, too. Peters and Lambert also argue for removing barriers to voluntary labeling. The FDA bars labeling food as "GM free" because it considers the term misleading. All food contains traces of GM products, and conventional food comes from plants genetically modified by traditional plant breeding. Furthermore, the FDA requires that a label advertising the absence of GM ingredients (in suitable terminology) also state that non-GM food is not more healthful than GM food. Peters and Lambert contend that these restrictions are unjustified and should be removed to encourage voluntary labeling.

In chapter 10, Clark Wolf evaluates various types of labeling regulations to see whether any constitutes an improvement of the status quo. The FDA, state government, and federal government all make labeling decisions. Are they permitted or obliged to require mandatory or voluntary labeling? Mandatory labeling may drive GM food from the market, reducing consumer choice among food products and increasing the risk of famine in parts of the globe. Regulation should target real risks and not perceived risks. It should advance the public's interests rather than its desires. Labeling policy should use the least restrictive means of achieving democratic goals. Wolf favors FDA supervised voluntary labeling. The public will pay a premium for non-GM food, and so voluntary labeling brings a benefit. It is more efficient to label non-GM food than to label GM food because most foods contain GM ingredients.

Carl Cranor considers in chapter 11 whether institutional considerations support mandatory labeling, not just voluntary labeling. Science is more likely to fail to detect a risk than to incorrectly assert that a risk exists. Showing that GM food is risk-free is hard. Relying on GM food may have adverse environmental consequences because genetic modifications spread to wild plants. Ecology and genetics are new sciences, and they have limited understanding of their subjects. According to the 2002 National Research Council report *Environmental Effects of Transgenic Plants*, the U.S. Department of Agriculture has inadequate procedures for protecting the environment from risks that GM crops impose. These risks have many features that may make them unacceptable to the public. For instance, these risks arise from technology and are borne by people who did not create them and do not even know of their existence. Consumers may reasonably discourage such risks. The disclosures they seek have precedents. Disclosures concerning the history rather than the current state of a product are mandatory in real estate transactions. They do not prevent trade. Labeling GM food will not hobble agbiotechnology.

Paul Weirich considers whether mandatory labeling is an appropriate means of reducing risk in chapter 12. What proofs of risk reduction justify a regulation? Should a government regulate to reduce a risk only after scientific investigation yields a full and conclusive assessment of the risk? Democratic principles support regulation to reduce imperfectly understood risks. Citizens may reasonably seek relief from those risks. A regulatory agency making decisions on behalf of the public should blend expert opinion and the public's interests to justify steps in areas such as food labeling. The best available information may not settle a risk's size. Aversion to the risk is still sensible. A regulatory agency may justifiably require that food labels provide information that allows consumers to avoid imperfectly understood risks.

Taken together, the essays evaluate labeling GM food as a middle course that keeps the door open to agbiotechnology but gives consumers some control over its direction. With the introduction of new foods from GM animals, consumer desires for choice are apt to intensify. Having thoughtful labeling policies will become more important than ever.

# Labeling Genetically Modified Food

# 1

# A Scientific Perspective on Labeling Genetically Modified Food

*Michael W. Pariza*

## WHAT IS "FOOD"?

We should begin by defining the word "food." On one level this may seem obvious to the point of being trivial. For example, one might say the term "food" refers to articles used for food or drink by man or other animals. In other words, if humans or animals think that something is a food, then by definition that something is a food. While this may seem rhetorical, the sort of definition that seeks to explain everything while in fact explaining nothing, it does, nonetheless, have important utility. It is also how the U.S. federal government defines food (the United States Federal Food, Drug, and Cosmetic Act, 21 USC 321[f][2006]). (For completeness, the government also includes in its definition synthetic and naturally occurring ingredients that are intentionally added to food, as well as chewing gum, perhaps mentioned specifically not only because we ingest the ingredients in gum but also because anxious students have been known to swallow their gum rather than admit to breaking the rule against chewing in class.) For the purposes of this discussion, I refer to this as the *legal* definition of food.

Another definition of "food" flows from taking a reductionist approach: Food is defined as a complex mixture of chemicals that humans and animals consume to nourish and sustain life. This, which I refer to as the *reductionist* definition, is clearly unappetizing but nonetheless does a good job of capturing the material nature of food without the transcendental associations of culture and tradition. Nutritionists and toxicologists will quickly add that the complex chemical mixtures that typically compose any given food may consist of chemicals that under some circumstances may be beneficial, benign, or possibly even harmful (Pariza 1996)—another good reason to follow mother's advice to eat a variety of foods.

Then there is what one might call the *biologist's* definition of food. In this case, food is defined as the products of other living organisms, each of which has its own unique evolution, and none of which has evolved simply to provide humans with nourishment (except of course human breast milk, the sole purpose of which is to nourish human infants). However, because of human intervention, our most common foods are in fact derived from plants and animals with which we have developed highly dependent synergistic relationships. In many cases, we truly need each other—they provide us with food, and we ensure their survival. (It is assumed in this definition that the products of other living organisms will also contain water, minerals, and other inorganic substances that may be essential for life but acquired by a plant or animal from external sources.)

We should also consider the meaning of adjectives that are often applied to food, for example, "novel," "biotechnology derived," and "genetically modified." Throughout human history, there have been examples of foods that were novel in one region or locale while simultaneously being dietary staples in another part of the world. In the years following the voyages of Columbus, a large number of foods were introduced to Europe, including corn, potatoes, tomatoes, squash, and cacao (chocolate) (Goldblith 1992). These were well-known foods to Native Americans but novel for Europeans, and Europeans often reacted with an understandable degree of caution. For example, many Europeans were initially concerned about toxicity from tomatoes because the tomato plant resembles the deadly European nightshade plant; fortunately for the evolution of Italian cuisine, this concern has been put to rest. Over time, these "novel" foods have been modified by agricultural practices that include very traditional techniques (e.g., plant breeding, cross-breeding, selection) as well as very modern genetic modification technologies that have become available only in recent years (e.g., recombinant DNA) (International Food Biotechnology Council 1990). Food processing technologies have also substantially changed these foods, for example, the invention of milk chocolate. Accordingly, it is appropriate to consider the terms "novel," "biotechnology," and "genetically modified" (GM) in the context of a continuum of agricultural and food production evolution that began thousands of years ago.

## FOOD AND FOOD LABELS

The focus of this book should be considered in the context of the foregoing. The legal definition is found in the statute that authorizes the Food and Drug Administration (FDA) to regulate food and food ingredients in the United States (the United States Federal Food, Drug, and Cosmetic Act, 21 USC 321[f][2006]). As such, this definition is broad and easily interpreted: Food is whatever consumers think it is. The FDA is thereby given sweeping authority to regulate the safety and quality of whatever it is that consumers may eat or drink. But while this definition tells FDA what to regulate, it does not give guidance on how to regulate. That is where the reductionist and biologist's definitions of food come into play, because without a scientifically based understanding of the nature of food, it would not be possible to develop a rational regulatory framework (International Food Biotechnology Council 1990).

The reductionist definition tells us that, whatever else, food is in the end a complex mixture of chemical substances (Pariza 1996; International Food Biotechnology Council 1990). Most consumers are aware of food labels, but these describe only the added ingredients. The ingredients may be defined substances, for example, ordinary table salt, which is a single chemical (sodium chloride); defined mixtures of substances, for example, high-fructose corn syrup, which consists of the natural sugars glucose and fructose; or whole foods, for example, milk, which contains many naturally occurring chemicals that are produced within the body of a cow or passed through from the food the cow ate. The ingredient list might also include natural cheese, which contains all the naturally occurring chemicals in milk plus the bacteria used to manufacture the cheese as well as metabolites naturally produced by those bacteria during the manufacturing process. And if the cheese is cooked, for example, on a pizza, then an entirely new mixture of chemicals will be produced as a result of heat-catalyzed reactions between various naturally occurring chemical components of the cheese. In other words, the food label lists only a very tiny fraction of the actual food constituents (Pariza 1996; International Food Biotechnology Council 1990).

The FDA's position is that the information on a food label should be useful to the consumer with regard to nutrition and health decisions. Hence, added salt is listed because some consumers should control their salt intake; added sugar is listed because some consumers need to control their sugar intake and sugar can also contribute to dental caries; total fat is listed because many consumers want to control their fat intake; trans-fat is now listed because it can contribute to elevated blood cholesterol; peanuts and other potential food allergens are listed to warn consumers with severe food allergies to these ingredients; and so forth. However, the innumerable minor naturally occurring or heat-derived chemical constituents that are found in virtually all foods are not listed because the information would be of no practical use to a typical consumer and, of course, because there is insufficient room on a typical label for such an array of declarations. If, however, qualified experts were to conclude that a particular naturally occurring or heat-generated substance might indeed pose a health risk for a segment of the population, then the FDA might indeed mandate that the substance be identified on the label.

The biologist's definition reminds us that humans are part of a larger order. Virtually all of the food on Earth comes as a result of the sun providing energy to green plants, which in turn use that energy to split water and carbon dioxide to produce glucose and molecular oxygen through photosynthesis. The resulting glucose is then metabolized in highly controlled and efficient fashion via the action of specific enzymes to ultimately produce proteins, lipids, nucleic acids, complex carbohydrates, and so forth. Animals eat the plants (as well as other animals that have eaten plants) to acquire needed nutrients, so in a very real sense our existence is very nearly totally dependent on sunlight and green plants.

But what sorts of green plants do we eat? The answer is, very few (International Food Biotechnology Council 1990). From a human perspective, the vast majority of green plant species are inedible for myriad reasons, including unacceptable taste, palatability, nutritional quality, and/or toxicity. The properties that make them unacceptable for our tables are precisely the same qualities that make it possible for plants to survive in hostile natural environments where they are in constant danger of being eaten by animals that can range in size from elephants to mites, as well as a legion of ubiquitous microbial pathogens and parasites, including viruses, bacteria, molds, higher fungi, protozoa, and so on. To protect themselves from predators and pests, successful plant species have evolved numerous defense mechanisms that include synthesizing naturally occurring pesticidal chemicals (Ames et al. 1990), and it is largely these defense mechanisms that make the plants unappetizing (which, of course, from the plant's "perspective," is entirely the point). One can easily visualize the success of this defensive strategy by taking a summer stroll in a woods: Green plants dominate the scene even in the presence of insects (that one can see) and microbes (that are invisible to the naked eye).

So, again, what sorts of green plants do we eat? In fact, we eat only a very few plant varieties that have been selected, cultivated, and genetically altered with a range of traditional and modern agricultural practices and technologies to match the qualities of taste, palatability, nutritional properties, and lack of toxicity that humans value. In fact, the wild relatives of the plants one typically finds in a supermarket would be largely unrecognizable (International Food Biotechnology Council 1990). Most of the plant species that we regularly eat are substantially weakened genetically, and they survive only because of constant human intervention and protection. In a remarkable example of symbiosis driven by human imagination, we protect and nurture our crop plants, and they provide us with food.

It should come as no surprise that the organoleptic qualities in a crop plant that we prize are also prized by other species. Accordingly, our success in crop improvement creates a dilemma: How do we protect the plants we've selected from being devoured by pests and pathogens? Natural armaments, the naturally occurring pesticides and physical attributes that protect plants in a natural environment, are no longer effective for many food crops because these protective mechanisms are typically linked within common biochemical pathways that were inevitably reduced as we selected the traits we desired (International Food Biotechnology Council 1990). Accordingly, we need alternative strategies, and the principal strategies in current use include synthetic pesticides, crop and integrated pest management programs, and biotechnology-based solutions.

## ASSURING THE SAFETY OF GM FOODS

The philosophical and legal debate surrounding GM crops includes two issues that are amenable to scientific inquiry: concern for the environment, and concern for human health (the creation of GM animals triggers an additional concern regarding animal welfare, which is beyond the scope of this discussion).

Environmental concerns associated with GM crops have been addressed in recent reports (Council for Agricultural Science and Technology 2003; Committee on Environmental Impacts 2002). Experts in agricultural science generally agree that that there is no one best way to protect all crops from all pests and diseases. Each approach, for example, the use of synthetic pesticides, crop and integrated pest management systems, and biotechnology-based solutions, has limitations. Rather than using a single approach, the consensus recommendation is to utilize all available technologies in a manner that is consistent with safety and sound land management. From an ecological standpoint, it may even be counterproductive to favor any one approach at the expense of others, unless the basis for doing so is scientifically sound. For example, a report by the National Research Council of the National Academy of Sciences (Committee on Environmental Impacts 2002) contained the following observation:

> *The committee recognizes that in any attempt to mitigate environmental risk there is a need to be mindful of the fact that avoiding one risk can sometimes inadvertently cause another greater risk.* [emphasis in original] For example, a regulation that discouraged research on pest-specific, plant-produced compounds could in some cases lead to continued use of environmentally disruptive synthetic pesticide. (page 7)

Human health concerns have also been addressed, and a framework for evaluating the safety of GM food has been established (International Food Biotechnology Council 1990; Kessler et al. 1992). In essence, the novel food is compared with its traditional counterpart via series of carefully crafted questions, including the requirement that the intake of any new constituents be shown to present no safety concerns. This framework, based on established scientific principles, is accepted by the FDA and has been successfully used in the United States for more than a decade to evaluate the safety of GM foods.

## THE PHILOSOPHICAL AND LEGAL DEBATE AND THE LIMITS OF SCIENCE

As discussed above, the FDA's position is that the information on a food label should be useful to the consumer with regard to nutrition and health decisions. This position does not recognize a consumer's "right to know" simply for the sake "knowing," nor does it recognize a manufacturer's "right to inform" simply for the sake of "informing." Rather, the FDA's position appears to be aimed at protecting the integrity of the food label against being used to advocate for or against whatever food controversies happen to be fashionable at a given time.

This policy is at the heart of the philosophical and legal debate that is the basis for this book. Science can help sort out the environmental and human health safety issues, but it cannot settle philosophical and legal disputes that deal with matters beyond what can be tested using the scientific method. In this regard, it is worth recounting the experience regarding the introduction of a GM tomato product in the United Kingdom by J. Sainsbury and Safeway stores, as documented by the National Centre for Biotechnology Education at the University of Reading, which reads, in part:

> In February 1996, J. Sainsbury and Safeway Stores in the United Kingdom introduced Europe's first genetically-modified food product. The modified tomatoes were grown in America, but they had been developed in the UK at Nottingham University and at Zeneca Seeds, based in Bracknell. The two retailers did everything you might expect of responsible firms: they labelled the tins very clearly, even though there was no legal obligation at that time for them to do so; they made sure that an alternative non-GM product was always available alongside the modified one; and additional information was available in leaflets in the stores and a telephone

help line. Rather than avoiding publicity they encouraged it, in newspaper and magazine articles and in radio interviews. The Consumers' Association applauded this approach and sales of the product were brisk. If there can be said to be a "right" way of introducing such a product, this came pretty close.

Three years later, almost to the day, everything changed. Friends of the Earth held a press conference at the House of Commons, highlighting the preliminary findings of a small-scale study in which rats had been fed GM potatoes. Arpad Pusztai's research had not at the time been published, and the memorandum presented by his supporters was principally concerned with securing Pusztai's right to publish his findings, and highlighting the apparently heavy-handed way in which he had been treated by his employers.

Pusztai's work had been reported by a "Panorama" television programme the previous summer but had gained little attention. Now, however, the opponents of GM food swung into action with a carefully-orchestrated campaign whose effectiveness shocked even its organisers. For several weeks in February 1999, each day almost all UK newspapers carried many pages of GM-related articles. TV reports showed maize being dumped at the gates to Downing Street, protesters dressed as animals, scientists munching on tomatoes and the Prime Minister depicted as Frankenstein's monster. This media circus was fuelled by press releases, publicity stunts, claims and counter claims from both sides of an increasingly-polarised debate. The most recent (2002) Eurobarometer survey conducted by the European Commission noted that 1999 was indeed the turning point in Europeans' attitudes to GM food.

Arpad Pusztai's rat-feeding study was eventually published in *The Lancet*. A Royal Society report in May 1999 criticised the design and conduct of the study and its statistical analysis. Even Pasztai's original supporters had noted the study's apparent shortcomings in their memorandum, and had urged that more research was needed (although they admitted that they had not actually seen Pusztai's work and when, subsequently, it was published and subjected to public scrutiny, several of them distanced themselves from it). To be fair to Pusztai, the study was only preliminary and was never intended to be published. Pusztai, however, has staunchly defended his original work against those who thought it flawed. Others have suggested that Pusztai was the victim of a wide-ranging pro-GM conspiracy.

Whatever the truth of these particular claims, today there is hardly any GM food in European shops, largely because of the success of the anti-GM campaign, triggered (in the UK at least) by the Pusztai work. …

… Reiss and Straughan, in their study of the science and ethics of genetic engineering, point out that arguments against GM generally fall into two categories: those that regard the technology as fundamentally wrong in principle (i.e., intrinsically wrong); and those that focus on the potential consequences of its application (i.e., its extrinsic properties). Note that similar sorts of reasoning also apply to some who speak in favour of the technology.

With those who object to (or support) GM on principle (e.g., by arguing that the process is morally objectionable (or desirable)) there is no debate to be had. Their moral views should be respected and that is that.

For those whose arguments are based on extrinsic concerns, however, the choice is less easy. Essentially they are making a prediction about what might happen if a particular technology is used. The evidence, scientific and non-scientific, is difficult to assess, even for those with a knowledge and understanding of the subject. After considering all the arguments, assuming a person does not decide to reject the technology entirely, what sort of GM crops might they accept, and which ones might be rejected? What regulations and safeguards should be put in place, and how should these be policed? How, if at all, can it be ensured that farmers of all types both here and overseas are not disadvantaged, and that vulnerable ecosystems are protected? [Quoted with permission from the National Centre for Biotechnology Education, University of Reading.]

Rarely has the essence of a societal debate been captured with such clarity. There is indeed a raging philosophical and legal debate. Both sides attempt to use science, but the scientific method is as delicate as it is sharp, and it is easily misused and manipulated. Scientists must guard against that. However, it is also incumbent upon philosophical and legal scholars to guard against the misuse of their disciplines in this crucially important debate.

## ACKNOWLEDGMENT

Thanks to Dean Madden at the University of Reading for permission to quote material on genetically modified food that Dr. Madden wrote and posted on the Web site of the National Centre for Biotechnology Education (www.ncbe.reading.ac.uk/NCBE/GMFOOD/menu.html).

## References

Ames, B. N., Profet, M., and Gold, L. S. 1990. Dietary pesticides are 99.99% all natural. *Proc. Natl. Acad. Sci. U.S.A.*, 87: 7777–7781.

Committee on Environmental Impacts Associated with Commercialization of Transgenic Plants, National Research Council. 2002. *Environmental Effects of Transgenic Plants: The Scope and Adequacy of Regulation*. National Academy Press, Washington, D.C.

Council for Agricultural Science and Technology. 2003. *Management of Pest Resistance: Strategies Using Crop Management, Biotechnology, and Pesticides*. Ames, IA: Council for Agricultural Science and Technology.

Goldblith, S. A. 1992. The legacy of Columbus, with particular reference to foods. *Food Technol.* 46: 62–85.

International Food Biotechnology Council. 1990. Biotechnologies and food: Assuring the safety of foods produced by genetic manipulation. *Regul. Toxicol. Pharmacol.* 12: S1–S196.

Kessler, D. A., Taylor, M. R., and Maryanski, J. H. 1992. The safety of foods developed by biotechnology. *Science* 256: 1747–1749, 1832.

National Centre for Biotechnology Education, University of Reading, Reading, UK. http://www.ncbe.reading.ac.uk/NCBE/GMFOOD/menu.html.

Pariza, M. W. 1996. Toxic substances in foods. Chapter 57 in *Present Knowledge in Nutrition*, 7th ed., ed. E. E. Ziegler and L. J. Filer, Jr. Washington, D.C.: ILSI Press, 563–573.

United States Federal Food, Drug, and Cosmetic Act, 21 U.S.C. 321(f)(2006).

2

# Genetically Modified Organisms for Agricultural Food Production

## The Extent of the Art and the State of the Science

*R. Michael Roberts*

### THE SCOPE OF GMOs

Anyone traveling through the Midwest agricultural belt of the United States (and presumably also comparable regions of Brazil and Argentina) in the growing season cannot be but impressed by the lush, uniform greenness of seemingly endless fields of soybean, a crop characteristically alternating with corn (*Zea mays*) on a regular two-year rotational basis. The chances are that this uniformity is in large part because the majority of soybeans on view have been derived from cultivars engineered to carry a gene that allow them to survive spraying with the broad-spectrum herbicide glyphosate, sold most commonly by the Monsanto Corporation as Roundup. Whether this change in agricultural practice provides overall benefits to society and the environment remains controversial. Nevertheless, genetically modified (GM) soybeans accounted for 55% of the world's and more than 80% of the U.S.'s soybean crop in 2003 (Nitta, 2004), and this transformation from ordinary cultivars occurred within a span of 10 years.

Examination of the corn growing in adjacent fields might be equally revealing. The chances are that these plants are pest protected and contain genes from the soil bacterium *Bacillus thuringiensis* (Bt) (National Research Council, 2000). The Bt gene encodes a protein harmless to animals, including humans, but when ingested by an insect pest stops the gut from functioning properly, thereby causing the targeted insect, usually its larva feeding on the crop, to die quickly. Bt was originally sold in the form of an insecticide spray in the 1950s, and its use is approved for organic farming because it is a so-called natural insecticide. At least 60 forms of Bt have been identified, each with their own range of insect specificities. Accordingly, plants have been engineered to resist the pests that most commonly plague them, in the case of North American corn, the larvae of the European corn borer. Well more than half of the corn crop in the United States is genetically modified in some manner, to provide pest control, herbicide resistance, or some other trait that improves yield or value of the crop.

Present estimates indicate that GM crops were grown on more than 100 million hectares in at least 25 countries in 2005, although six countries (the United States, Argentina, Canada, Brazil, China, and South Africa) led the way. I open this short essay with this information to impress the reader that GM crops are now part of global agriculture (McHughen, 2000; Winston, 2002). They first hit the fields in 1996, but now have a value exceeding $5 billion annually. They have gained wide acceptance by farmers, Agribusiness, and even consumers in large parts of the world and are knocking on the door of those countries that are avowed to keep them at bay. The battle is over; GMOs (genetically modified organisms) are here to stay. The real question is what products will appear next and what influence they will have on society, the global economy, and the environment in which we live.

## WHAT ARE GMOs?

Any description of genetic modification is complicated by semantics. Different people mean different things when they use the term, and the various national regulatory agencies have themselves failed to provide consistent definitions. Mankind has been manipulating, that is, genetically modifying, the genomes of plants and animals since the beginnings of agriculture. The earliest farmers selected grasses that did not shed their seeds and that had the biggest ears, and wild cattle and hogs that were most docile and easy to herd. Selective breeding generated cultivars and breeds totally unfit to survive in the wild but well adapted to the farmers' needs and practices. These empirical approaches to exploiting natural variation, particularly crosses between only distantly related organisms, involve the uncontrolled shuffling of thousands of genes.

Genetic modification in modern parlance, however, is generally regarded as different from these older practices of selective plant breeding and animal husbandry and involves the use of recombinant DNA technology (McHughen, 2000; Winston, 2002). This cultural revolution in science began in the 1970s with the discovery that DNA, and hence the genes it encodes, could be cut up into pieces, copied, and rejoined in new combinations. A GMO is usually one that has been deliberately created to contain a piece of "foreign" DNA, usually a full-length "foreign" gene incorporated in its genome. This inserted DNA, which is present in the host cell's chromosomes, will reproduce itself in step with the all the resident genes that surround it and will be passed to subsequent generations as the organism reproduces. In general, the foreign DNA represents a gene responsible for creating a perceived "desirable" trait, for example, resistance to glyphosate or the presence of Bt, not normally associated with the organism and often derived from the genome of another organism with which the recipient would not normally interbreed. Such a "lateral" transfer of genetic information across species barriers occurs naturally outside the laboratory and is quite common in bacteria but rare, although not unknown, in plants and animals. In the last 25 years, scientists have learned how to speed up this process and render it precise and reproducible. The addition of novel genes to crop plants, animals, and even their insect pests is now feasible for just about any gene. It is now mainstream science.

An alternative form of genetic modification is gene ablation, where an undesirable trait is removed from the genome of an organism rather than added. For example, if the genes encoding potential allergens, for instance, those in nuts, can be identified, it is theoretically possible to eliminate them from cultivars. In the case of animals, thousands of gene "knockouts" have been performed in the mouse, including the one for the gene encoding the prion protein. Eliminating such a gene in cattle, sheep, and goats, providing they can survive without it, could prevent any risk of prion disease ("mad cow" disease) arising from consumption of meat (National Research Council, 2002).

Rather than describe the various techniques that allow such genetic modifications to be performed in the laboratory—there simply is not enough space to do so in a short essay of this kind—the reader should be aware that the process, if not routine, is relatively straightforward. It is now possible to create a GM crop plant within months, cross the modified plant with an elite, high-producing cultivar, and select for lines suitable for field testing within a few growing seasons (McHughen, 2000). The process does not end there, however. Before a new cultivar, whether or not it is a GMO, can be sold to farmers or gardeners, the new variety has to be shown to show some new trait that sets it apart from other cultivars, must produce a crop where all the plants look alike, and must exhibit genetic stability, that is, not lose its characteristics over time. Finally, sufficient seed must be produced to meet sales demand.

The barrier for GMOs is set even higher than for ordinary cultivars. The regulatory bodies require that any novel product likely to enter the food chain be safe for consumption and that, as far as possible, the risk of the integrated gene spreading to other cultivars and even wild populations of related species is minimized (McHughen, 2000; Winston, 2002). Despite these barriers, the numbers of patents, registrations, and approvals relating to GM crop plants continues to rise every year. How quickly these new GMOs become commercial products will depend upon their perceived safety, their acceptance by consumers, the reach of the regulators, and above all, whether the product has the potential to earn profit for the companies that develop it.

## THE SPECIAL CASE OF GM ANIMALS

Although many kinds of GM plant have been approved for commercial use, this is not the case for animals (National Research Council, 2002). The U.S. Food and Drug Administration (FDA) is still deciding how to regulate agricultural products from animals that have been genetically engineered or produced by cloning, both controversial technologies that the public finds even more unsettling than the genetic modification of plants. There are some good reasons for concerns. Farm animals such as cattle, sheep, and goats have generally been genetically modified not to be better producers of milk, meat, and fiber but to manufacture pharmaceuticals, such as blood clotting medicines, in their milk. Pigs with livers, kidneys, and hearts specially tailored not to be rejected in organ donation are also on the horizon. Should these animals be processed for food? Probably not, since residues of drugs and anesthetics in the carcasses of such animals would make them too risky to eat.

But what about GM animals that have been raised primarily for agricultural purposes, such as cows producing nutritionally enhanced milk or chickens releasing less phosphate in their droppings? These cases would appear to be analogous to those of GM crops and will likely be evaluated individually for safety and environmental concerns.

Cloning, to some, is the scariest of the new technologies in animal agriculture, although we have been cloning many kinds of plants, for example, by cuttings and grafting and by using bulbs, corms, and rhizomes, for thousands of years. A major purpose of animal cloning is to "copy" agriculturally valuable, elite animals, and cloning is also likely to have a role in preserving rare breeds from extinction. Clones, by definition, are genetic copies, so the milk and meat from such animals is unlikely to be different from an animal produced by normal reproduction. Until quite recently producers of cloned livestock, operating under a voluntary moratorium, have been obliged to discard milk, meat, and other products from cloned animals (Rudenko & Matheson, 2007). In late 2006, however, the United States Food and Drug Administration (FDA), after a full review of the evidence, decided that there were no hazards to consuming meat and milk from cattle, swine, and goat clones and that the food was as safe as that from animals derived by conventional means. It will be interesting to see what labeling strategy, if any, will be applied to the products once these are approved for sale. It should be noted that, even though cloning through so-called somatic cell nuclear transfer is only a decade old, beginning with Dolly the sheep, scientists have been "splitting" embryos to create twins and even quadruplets for more than 20 years, and the milk and meat products from such cattle safely entered the food chain without special regulation or labeling (National Research Council, 2002).

## GMOs FOR HUMAN CONSUMPTION

Although a large proportion of GM corn, soybean, and other crops is fed to animals (and, recently, to generate biofuels), a significant portion also finds its way to the shelves of supermarkets, much of it in the form of packaged, highly processed products. Moreover, a range of new and second-generation GMO plants are being developed that might offer some attractive benefits to consumers in terms of nutrition, for example, rice enriched with vitamin A, oils with reduced saturated fats, and other properties that were sadly lacking in the pest- and herbicide-resistant first generation of GM crops.

Deciding how such foods are classified, sorted, and labeled, should such controls be felt necessary, will be an increasing challenge to governments, regulatory bodies, and consumers. Let us consider a few examples that already confront us. Is canola oil recovered from a canola cultivar resistant to Roundup any different from the oil of a nonmodified plant? Probably not, but is that sufficient to allay fears of those who regard GMOs with such suspicion? It contains none of the feared, but harmless, Bt pesticide. Similarly, high-fructose corn syrup, a major value-added product from both standard and GM varieties of corn, is a carbohydrate and hence is free of the proteins encoded by the genes that provide pest and herbicide resistance to corn. And is the meat and fiber from animals that eat GMOs somehow tainted with the residues of genetic modification? Again, the answer is almost certainly no.

One particular challenge to regulators will be how to deal with food products where a native gene, for example, one encoding an allergen, has been deleted from the genome rather than a foreign gene added. Such plants are, without doubt, genetically modified, but is the lack of a gene important to the consumer?

Finally, why are GMOs excluded from being blessed with the designation "organic" even if they are grown according to the organic handbook? Advocates of GMO foods argue that we should not be regulating the technology itself, as the European Union and some groups of consumers demand. Instead, they argue, the primary concern should be the safety of the products derived from the GMOs and whether these products are different in composition than those from the unmodified relative. However, consumers often rightfully feel that they should be informed whether the coffee they drink, the running shoes they wear, and, of course, the food they eat are produced according to practices of which they approve. The challenge of any open discussion on labeling of products derived from GMOs is to address such dilemmas.

## DETECTION AND MEASUREMENT OF GMOs

The European Union requires that if food stuffs contain more than 1% of their ingredients from GMOs, they must be labeled (see chapter 4). Japan has zero tolerance for GMO "contaminants." Organic farmers must also be assured that their seed is GMO free. To fill this demand for accurate analysis, companies have developed a range of methods to detect small amounts the foreign DNA or the protein the gene encodes in a large excess of unmodified product. Genetically modified organism testing has become a major industry in its own right, with many commercial laboratories offering services.

There is a wide range of GMO testing protocols, each with its own set of advantages and disadvantages. Perhaps the simplest are bioassays to detect herbicide resistance genes in samples of seeds. Here the approach is to simply germinate the seeds in presence of the herbicide. The GMO seed will germinate normally; the nonresistant strains will die. Provided that the sampling reflects the makeup of the mixture of seeds analyzed, that is, a proper statistical model is applied to the test, the extent of GMO contamination can be easily and quite cheaply assessed.

More sophisticated assays measure the presence of the suspect protein by means of antibody reagents. These tests are analogous to many used in medicine to measure the levels of insulin or other hormones present in small amounts in blood. They can be extremely sensitive and highly specific and can provide accurate measurement of the relative amount of protein present in a sample that might contain both GM and "native" constituents. "Dipsticks" and "strip tests" are less accurate than some laboratory methods, but since they are in kit form, they are quick and require little sophisticated training by the user. The disadvantage of immunoassays is that they often do not work if the protein has been denatured by cooking or other forms of processing.

A third approach is to detect the gene itself by the polymerase chain reaction (PCR) procedure. This method can be extraordinarily sensitive and detect minute traces of product, since the tests involve amplification of the DNA composing part of the foreign gene. PCR techniques are also used widely in diagnostic medicine. The PCR approach is expensive, demands sophisticated equipment and training, and presently cannot be used on the farm or at the grain elevator.

## CONCLUSIONS

Genetically modified crop plants are now a mainstream part of the agricultural landscape in many parts of the world. Genetically modified animals have not yet made a contribution to animal agriculture (see chapter 5) but may do soon in the form of growth-enhanced salmon. The trend toward increased use of GMOs in plant and eventually animal agriculture will undoubtedly continue as long as there are markets for the products. Since the technologies used to add or remove genes from plant cultivars are now almost routine, there seems little that can be done to restrict the spread of GMOs and their food products, even if some legislative bodies consider such actions desirable. The challenge will be to ensure that regulatory standards in terms of food safety and potential for environmental harm are maintained globally. There is no equivalent of a nuclear non-proliferation treaty in the GMO world. On the other hand, because the sensitivity and sophistication of detection methods to assess the presence of foreign genes and their products in agricultural commodities and processed foods will continue to improve, the consumer can, in theory, be protected from the involuntary consumption of such materials by appropriate labeling regulations backed up by rigorous product analysis. The questions of whether such labeling is either appropriate or necessary, how much "contamination" of food products should be allowable, and what to do when the original organism was genetically modified but the food itself bears no lingering residue of the trait remain the topic of books such as this one.

## References

McHughen, A. 2000. *Pandora's Picnic Basket: The Potential and Hazards of Genetically Modified Foods*. New York: Oxford University Press.

National Research Council. 2000. *Genetically Modified Pest-Protected Plants: Science and Regulation*. Washington, D.C.: National Academy Press.

National Research Council. 2002. *Animal Biotechnology: Science-Based Concerns*. Washington, D.C.: National Academy Press.

Nitta, Itaru. 2004. *Genetically Modified Soybeans and the Patent System.* TED Case Studies No. 769. Available at http://www.american.edu/ted/soybean-patent.htm.

Rudenko, L., Matheson, J. C. 2007. *The US FDA and animal cloning: risk and regulatory approach.* 67(1): 198–206.

Winston, M. L. 2002. *Travels in the Genetically Modified Zone.* Cambridge, Mass.: Harvard University Press.

# 3

# Biotechnology and the Food Label

*Fred H. Degnan*

Consider the food label. It defies simple classification. It comes in numerous sizes and shapes, from cereal boxes to chewing gum wrappers, from cylindrical cans to fancifully shaped plastic bottles. It can bear a variety of messages delivered in a variety of ways and styles designed to attract the consumer's attention. It can advertise and promote. It can share recipes and dietary guidance. It can communicate public service announcements and disseminate the pictures of missing children. It can display garish likenesses of cartoon characters or exquisite portraits of professional athletes and idols. In short, it is the front line of product marketing and, thus, subject to fundamental yet idiosyncratic decisions as to what information, art work, form, and level of taste will attract consumer attention and induce a purchase.

Precisely because the food label can inspire and provide the critical stimulus for a food purchase, for over 100 years federal law has recognized that the government has an interest in ensuring that the food label serves to help consumers to make their food selections wisely. The precedent derived from this century of federal regulation sets the stage for a focused consideration of the issues involving the use of the food label to inform consumers about whether food is the product of biotechnology or contains ingredients that are produced through biotechnology. No current labeling issue is more controversial.

Part of the controversy derives from a lack of public trust about the short- and long-term safety of applying innovative genetic engineering techniques to the production of food. At work here, clearly, is the unease felt by many about the use of gene technology to develop and produce traditional foods—the use of technology in a context that touches our lives daily and personally. Headlines about gene therapy investigations gone awry and the cloning of one large animal species after another only serve to fuel this concern.

Nevertheless, the FDA has resisted the urgings of interested domestic and foreign parties alike to require, on a whole-scale basis, information on the food label about the genetically engineered status of a food or the ingredients it bears. This position is rooted in the FDA's science-based conclusion that, as a general rule, there is nothing inherently unsafe or mysterious about food biotechnology. Discontent with the FDA's scientific conclusion and labeling position has become commonplace. This chapter focuses on the legal rationality of the FDA's current labeling policies in light of decades of agency precedent and experience in policing the food label, and on agency use of its statutory authority to require information to appear on the label.

## REQUIRING INFORMATION TO APPEAR ON THE FOOD LABEL

### Five Essential Pieces of Information

For purposes of this analysis, there is a critical distinction between what the FDA can "require" to appear on the food label and what the FDA can "prohibit" from appearing on the food label. The FDA's fundamental food labeling authority is found in section 403 of the Food, Drug, and Cosmetic Act of 1938 (FDCA).[1] Parts of section 403 have been in effect for nearly 100 years—they were included in the first federal statute governing the regulation of food, the Pure Food and Drugs Act of 1906.[2] Section 8 of that early Act focused on the misbranding of food. The purpose of the section was to prevent consumers from being deceived by what sellers said about their product. This section proscribed the sale of any food labeled in a false or misleading manner. The section was not at all about "requiring" information to appear on the food label. Rather, its focus was (1) on whether information that manufacturers voluntarily placed on the food label was truthful and not misleading and, if not, (2) on "prohibiting" such information from appearing on the food label.

Section 403 itself first appeared with the passage in 1938 of the FDCA. In crafting section 403, Congress adopted essentially the identical language prohibiting false and misleading labeling as appeared in the 1906 statute. In so doing, Congress concluded that the general prohibition against false and misleading representations was meant to be comprehensive in character and recognized that "the labels of food... are not considered... to be the proper media for making any representations... which are not in accord with the facts."[3]

The framers of the 1938 Act also recognized that prohibiting false and misleading labeling, although critical to any regulatory function over the food label, did not fully equip or empower the FDA if the purpose of the label was to meaningfully inform consumers about food. Accordingly, they introduced a then novel concept in food labeling: Certain information was deemed so essential, so material, that the agency should have independent authority to *require* that this information be included on the label of *any* food. With this design in mind, the framers empowered the FDA to require four fundamental pieces of information: the identification of the ingredients used to fabricate the food; the prominent, clear declaration of the net weight of the contents of the food; the name and address of the manufacturer or responsible party involved in the marketing of the food; and a precise statement of the identity (the name) of the food.[4]

It does not take extensive reflection to grasp the essentiality of this required information. It is truly basic. That Congress sought to require only this information from every food manufacturer reveals a deliberate intent to limit the amount of information that could, categorically, be compelled to appear on the food label. These categorical requirements are distinct from other requirements in the current section

403 that relate to specific products, for example, products that contain any artificial flavoring, artificial coloring, or a chemical preservative.[5] These less universal requirements, for want of a better descriptor, were included in the original 1938 Act in an effort to correct a longstanding labeling abuse by preventing manufacturers from concealing damage or inferiority or to create a deceptive appearance to food.[6]

Fifty-two years passed before Congress added a fifth categorically required piece of information to appear on the food label: complete nutrition labeling. The requirement was the cornerstone of the Nutrition Labeling Education Act of 1990 (NLEA).[7] The framers of the 1990 legislation, like their counterparts in 1938, shared a comparable regard for the importance of the food label. The essence of the nutrition labeling requirement was the congressional desire that the food label convey meaningful nutrition information about foods in as simple, clear, and consistent a manner as possible. The goal of this NLEA-based reform was thus the same as the fundamental goal of section 403 when enacted in 1938: the communication of essential information to consumers.

Of course, the NLEA contained a number of other requirements related to representations that may appear on the food label. In each case, however, these requirements come into play only where a manufacturer wishes to make nutrition-related claims about its product. In enacting the NLEA, Congress was clearly of the mind that, in spite of the passage of more than 40 years, only nutrition labeling was of such essentiality as to merit an unconditional requirement comparable to the four other fundamental pieces of information required by the 1938 Congress.

That being the case, it is instructive to note that the mandatory nutrition labeling requirement of the NLEA was not designed to compel the disclosure of just routine information on food labels. Rather, the requirement was designed to require the disclosure of essential information that consumers need to choose foods wisely. To this end, consider that while NLEA specifies the nutrients for which information must be provided in nutrition labeling, the Act gives the FDA the authority to exclude any nutrient, regardless of its presumptive public health significance, from the declaration requirement when the agency finds that the information "is not *necessary* to assist consumers in maintaining healthy dietary practices"[8] (emphasis added).

In implementing the NLEA, the FDA was guided by this notion of essentiality. For example, in proposing its nutrition labeling regulations, the agency acknowledged that nutrient information previously required in labeling, but currently of no "pressing public health importance," needed to be excluded in order to ensure that critical messages about calories, cholesterol, fat, protein, and carbohydrates were meaningfully conveyed to consumers.[9] When the FDA adopted final rules excluding numerous declarations, it emphasized that "not all information related to maintaining healthy dietary practices can be included on the food label."[10] The agency went on to explain that not only would space constraints not allow for this, but also "the large amount of information would interfere with consumers' ability to use the information of the greatest public health significance."[11]

This emphasis on essentiality permeated other aspects of the FDA's NLEA-related rule making. Time and again, the agency acknowledged the difficult tradeoffs it was forced to make to ensure that essential information about foods was conveyed in a fashion that consumers could use. For example, noting that although the NLEA requires health claims to contain sufficient information to be understood by consumers in the context of the daily diet, the agency explained that "there is a limit to the amount and complexity of information that can be presented in a health claim" and that can reasonably be understood and used by consumers.[12] The FDA has reiterated these fundamental concerns in its present considerations regarding how best to use the food label as a vehicle for helping address the nation's obesity problem.[13]

In summary, express authority to require specific types of information to appear on the food label is limited under the FDCA to precise contours. The information that can be, for practical purposes, universally required is that which Congress perceived as essential to informed consumer decision making. Desirable but collateral information was not deemed by Congress either in 1938 or in 1990 to be of comparable status. The congressional goal was clearly to limit to essentials the express information that the FDA could require to appear on the food label. The value judgment implicit here was that consistent access to such fundamental information was the best way to employ the food label as a tool for informing the consumer. Of course, effective labeling policy also required, in conjunction with providing essential information, prohibiting false and misleading information. The FDA's long history of taking enforcement action against false and misleading labeling practices is well chronicled (Hutt and Merrill 1991: 36–75).

## The Authority to Require Additional "Material" Information

Section 201(n) of the FDCA was, like section 403, enacted in 1938.[14] It amplifies the general prohibition against false and misleading labeling that has been at the cornerstone of FDA regulation since 1906. Moreover, the section is an adjunct to the FDA's authority described above to require essential pieces of information to appear on the food label. In essence, this section gives the FDA the authority to require that the food label bear information in addition to the five fundamental elements if such information is necessary to prevent consumers from being misled. The section, however, is direct in its language and is limited to information that is, in fact, "material."

Relying on section 201(n), the FDA can require two types of information. Under one prong of the section, the FDA can require the disclosure of facts that are "material" to the "consequences" of consuming a food. The other prong is conditional in nature: It gives the FDA the authority to require a seller to reveal facts that are "material" in light of the seller's own representations about the food. In essence, this component of section 201(n) reflects a "tell the whole truth" standard that comes into play if a manufacturer opts to tell something about its product and, in the process, fails to provide all "material" information.

Consistent with the clear emphasis on "materiality," the FDA has, albeit rather infrequently, relied on section 201(n), in conjunction with section 403(a), to expressly require information to appear on the food label. Probably, the most notable and familiar example of the FDA's reliance on section 201(n) to require manufacturers to provide specific information is the agency's original (1973) nutrition labeling regulations.[15] There, the FDA reasoned that sections 403(a)(1) and 201(n) could justify imposing the requirement upon a manufacturer that added a nutrient to a food or that made a claim about the food's nutritional value to disclose important information about the nutritional value of the food. The FDA further reasoned that, in light of the claims made or clearly implied for such foods, the failure to provide nutrition information would constitute an omission of material fact. The implicit corollary to the agency's reasoning is important: Unless a manufacturer made a nutritional claim or took affirmative steps to fortify a food, the agency lacked authority to require a disclosure of information about the nutritional value of the food.

The other prong of section 201(n), the material consequences provision, recognizes that certain foods or food ingredients may present risks for some consumers that can, and should, be averted through affirmative labeling. To this end, the agency has required declarations identifying the presence of ingredients that may affect an existing disease condition. An example is the requirement to designate the ingredient gluten as derived from either corn or wheat to accommodate those with celiac disease who suffer serious reactions when exposed to wheat gluten.[16] Similarly, the FDA has required a special warning statement to appear on the label of protein products intended for use in weight reduction stating, in part, that very low calorie protein diets may cause serious illness or death.[17] Self-pressurized food containers must also bear label warnings against spraying in the eyes, puncturing, or incinerating (Hutt and Merrill 1991: 82).

In an interesting and instructive twist, the FDA in 1990, as part of broad agency activity to solidify aspects of its food labeling authority, proposed to make nutrition labeling mandatory on those foods that were meaningful sources of nutrients.[18] Once again, the agency relied on section 201(n) for its legal authority for such a requirement. The agency's rationale for applying section 201(n), however, differed from that employed in 1973. As noted above, the 1973 regulations were premised on the first prong of the section: the need to correct and balance a representation. The 1990 proposal was based on the second prong of the section: the need to inform consumers about the material, nutritional consequences of consuming or not consuming a particular food and the conclusion that basic nutritional information was, in fact, material with respect to the consequences of consuming particular foods. The agency's reliance on section 201(n) in this context was never tested. Three months after the agency proposed mandatory nutrition labeling, Congress passed the NLEA and amended section 403(a) to expressly make nutrition labeling a required component of the food label.

With respect to food, relevant to any consideration of section 201(n) is section 403(i), which requires that a food or food ingredient bear its common or usual name. The agency has employed section 201(n) in conjunction with section 403(i) to require that information regarding the source of the food or food ingredient be included in the common usual name in those cases where it is necessary to accurately identify the basic nature (including value) of the food. Applying this rationale, for example, the FDA concluded that the declaration of the source of protein in protein hydrolysates is necessary.[19] There are, however, a number of situations in which the agency has decided that source information is not a necessary part of the common or usual name of the food or is not material within the meaning of section 201(n).[20] What appears to have been key in the agency's decision in this area is whether the information is critical to identifying the basic nature of the food or the material consequences that may accompany its use.

The FDA's use of section 201(n) in requiring labeling statements regarding the consequences of consuming or using products other than foods is also instructive as to the rigor and purpose with which the agency has generally applied the provision. For example, the agency has required special patient labeling for oral contraceptives, hearing aids, and intrauterine devices in an effort to provide information about the consequences of the use of these products.[21] In the context of tampons, the FDA has required, again relying on section 201(n), labeling concerning the risk of toxic shock syndrome.[22] In this case, however, although the agency was asked to require ingredient labeling for tampons because of consumer concerns about the safety of some ingredients, the FDA was not aware of any data showing an association between any particular ingredient and any risk of health, including allergic reaction, sensitivity, and irritation. Due to this lack of scientific support, the agency refused to impose an ingredient labeling requirement.[23]

Even this rather rough and by no means exhaustive chronicle of the FDA's application of section 201(n) reveals the care that the FDA has generally employed in exercising the section. Not all material facts are required to be disclosed: It is only the omission of facts that are material *in light of the representations made* about a product or with respect to the consequences of a product's use that make its labeling misleading. Accordingly, the agency has routinely avoided requests for information or warning statements for ingredients that cause only mild or idiosyncratic responses. To not do so could overexpose consumers to mandated information and warnings and, in the process, crowd out more important information and blunt the intended and desired impact of such information.

The careful reader will note the use of the qualifier "generally" in the foregoing paragraph. Although the FDA's application of section 201 has been consistent over the years, there is at least one case in which the agency arguably has invoked section 201(n) to impose labeling requirements for which it is difficult to muster the degree of support that accompanies other applications of the section. The case involved irradiated foods.

Currently, foods that have been irradiated must bear a representation and a logo to that effect.[24] Irradiation is the only instance in which a processing technique has been required to be disclosed under section 201(n). The requirement was based not on safety concerns about irradiation but rather on the fact that irradiation could cause changes in flavor or shelf life of finished foods and that these changes could be significant and material in light of the consumer's perception of such foods as unprocessed.[25] Arguments can be assembled that the FDA may have focused on processing as material information in an effort to accommodate consumer concerns over the safety of irradiation. In so doing, the FDA may have stretched its traditional exercise of authority under section 201(n). In any event, in resolving the irradiation issue, the agency, sensitive to the conventional interpretation and use of section 201(n), went to rather great lengths to craft a rationale for the required disclosure that fell within the literal confines of the agency's traditional application of this section. In the process, the agency clearly sought to preserve the argument for future applications that any information required to appear on the food label should be truly material and essential.

Another case of the FDA's application of section 201(n) that, under close scrutiny, may be inconsistent with the agency's historical application of the section is its imposition of a boxed "information" statement on foods containing olestra.[26] Olestra is a no-calorie cooking oil. In 1996, the FDA approved olestra as a food additive for use in place of fats and oils in prepackaged, ready-to-eat savory snacks. Traditional toxicology studies showed that olestra is not nontoxic. Studies also showed that because olestra is lipophilic, it can absorb fat-soluble vitamins and nutrients. Research also revealed that because olestra is not absorbed, it passes through the gastrointestinal tract intact and thus, like many foods, has the potential, when eaten in large quantities, to cause gastrointestinal effects. The FDA concluded that the effects accompanying the consumption of olestra did not represent significant adverse health consequences.[27] With respect to the vitamin and nutrient issue, the agency concluded that it was possible to supplement foods containing olestra with vitamins A, D, E, and K in such a way as to compensate for the amounts that are not absorbed from the diet due to the action of olestra.[28] Nevertheless, in spite of these findings, the FDA relied on section 201(n) to impose "for an interim period" a labeling requirement alerting consumers about the potential for gastrointestinal effects. The agency reasoned that such a label would reassure consumers as to the source of any gastrointestinal effects as well as help consumers avoid unnecessary medical treatment. In addition, the FDA reasoned that in order to avoid possible consumer confusion about the reason why vitamins were added to olestra snacks, they should be provided with information about olestra's potential to inhibit the absorption of the fat-soluble vitamins. In spite of the restraint that characterizes the history of its use of section 201(n), the FDA relied on an unprecedented mechanism to convey this material information, a boxed "information statement" (see figure 3.1). Postapproval research on olestra revealed, among other things, not only that abdominal cramping is not a side effect but also that consumers perceive the boxed message not as an "information statement" but rather as a warning calling into question the safety of olestra.[29] As a result, the FDA, ultimately, formally withdrew the labeling requirement.[30]

In sum, for more than 60 years, the FDA has had in its enforcement and policymaking arsenal section 201(n)'s authority to require information to appear on the

**This Product Contains Olestra.** Olestra may cause abdominal cramping and loose stools. Olestra inhibits the absorption of some vitamins and other nutrients. Vitamins A,D,E, and K have been added.

**Figure 3.1.** The boxed "information statement" the FDA required to appear on products containing olestra between 1996 and 2003.

food label. The agency has exercised that authority sparingly, largely reserving its use for the disclosure of truly important, noncollateral and non-label-cluttering "material" information. The food label would be an entirely different entity from what it is today if the FDA had acted otherwise. For example, imagine the possible array of different types of information that could be "required" if the FDA deemed "public desire" for information an indicator of materiality. And, further, imagine the possibly unworkable contours, not to mention the size and space demands, of a label capable of accommodating such information.

## THE FDA's LABELING POLICY FOR FOOD PRODUCTS OF BIOTECHNOLOGY

### A Brief Background on Food Biotechnology

In the mid-1800s, Gregor Mendel demonstrated that plant characteristics could be inherited. Ever since, plant breeders have been trying to improve plants by altering their inherited characteristics, that is, their genetic makeup. Cross-breeding and hybridization served as the primary technique for accomplishing these genetic changes. In the breeding process, however, the potential always existed for unintended, and perhaps undesirable, traits to appear in addition to those intended. Breeders learned that time-consuming, additional breeding helped eliminate many, although not all, unintended traits. Breeders could then select the most desirable plant variety. As a result of these conventional genetic techniques of breeding and selection, many foods, now common in our diet, came into being. Hybrid corn, nectarines, and tangelos are examples of a few.

Every cell has a molecule of deoxyribonucleic acid (DNA) that encodes the genetic information for that cell. Each DNA molecule contains chromosomes and information for thousands of different genes. More than two decades ago, scientists discovered that they could "cut" a gene-size piece from a chromosome and attach that gene-containing piece to another cut chromosome (this led to the expression "gene splicing"). Today, by inserting a single gene into a plant, scientists are able to produce a plant with new characteristics. Because today's technologies—recombinant DNA technology—are so precise, they permit predictability about the qualities and traits to be conferred to a new plant variety. This is a meaningful improvement over traditional plant breeding and significantly decreases the potential for undesirable substances to be introduced into a plant. Another advantage of this new technology is efficiency: New plants can be developed much more rapidly than conventional hybrids.

In 1994, the FDA authorized the marketing of the first genetically engineered plant product—the Flavr Savr tomato. The tomato was engineered to contain a gene that delayed ripening. As a result, a tomato could stay longer on the vine to ripen to full flavor but remain firm during transport (after shipment, producers could induce ripening with ethylene gas, the natural ripening agent in tomatoes). This alteration avoided the common problem of overripening and bruising during shipment that accompanies commercially grown tomatoes. The FDA was not hasty in authorizing the introduction of the Flavr Savr tomato. The agency conducted a thorough scientific review of the manufacturer's data and information with respect to the tomato, its constituents, and the use of a kanamycin resistance marker gene in its production. The FDA held a public meeting of its Food Advisory Committee to examine the data collected with respect to the safety of the tomato and the procedures followed by the manufacturer in developing the tomato. Committee members agreed with the FDA that the scientific approach followed and the data collected were sound and that all questions regarding the safety of the tomato were addressed. Since the FDA's authorization of the Flavr Savr tomato, more than 60 other genetically engineered crop foods have been determined by the agency to be as safe as their conventional counterparts. These crops include soybeans resistant to the herbicide glyphosate, corn resistant to corn borers, and squash resistant to common viruses.

## Mandatory Labeling

In a policy statement issued on May 29, 1992, the FDA addressed the issue of how best, under the foregoing existing statutory rubrics, to label foods derived from plant biotechnology.[31] The agency concluded that consumers "must" be informed by appropriate labeling if a food derived from a new plant variety differs from its traditional counterpart in such a way that its common or usual name no longer applies or if a safety or usage issue exists to which consumers must be alerted. The agency explained its policy, in part, as follows:

> For example, if a tomato has had a peanut protein introduced into it and there is insufficient information to demonstrate that the introduced protein could not cause an allergic reaction in a susceptible population, a label declaration would be required to alert consumers who are allergic to peanuts so that they could avoid that tomato, even if its basic taste and texture remained unchanged. Such information would be a material fact whose omission may make the label of the tomato misleading....[32]

In response to the anticipated concern that foods or food ingredients developed using biotechnology techniques should be required to bear labeling to reveal that fact to consumers, the agency emphasized that it did not consider methods used in the development of a new plant variety, such as conventional hybridization, chemical or radiation-induced mutagenesis, protoplast fusion, embryo rescue, somoclonal variation, "or any other method," to be material information within the meaning of section 201(n).[33] The agency explained its belief that the new biotechnology techniques are simply extensions at the molecular level of traditional methods and are used to achieve the same goals as pursued with traditional plant breeding. The agency went on to emphasize that it was not aware of any information showing that foods derived by these new methods differ from other foods in any meaningful or uniform way or that as a class foods developed by the new techniques present any different or greater safety concern than foods developed by traditional plant breeding.

For most of the 1990s, this policy drew relatively little domestic attention. At the same time, however, the reaction to genetically modified foods overseas was markedly different, as public attitudes among consumers and regulators alike reflected skepticism about the new technology and genuine concern about its public health, environmental, and ethical implications. This climate led to the European Union's preparation and issuance in 1997 of a "directive" that outlined specific labeling requirements for food products of biotechnology.[34] Pressure for labeling and tighter safety controls for food products of biotechnology steadily mounted and, in mid-1999, rapidly escalated here in the United States. This escalation was prompted, in part, by publicity about the European Community's reluctance to accept vast amounts of corn and soy produced from modified seeds and exported to Europe by U.S. growers.

To address these growing concerns, the FDA in late 1999 held three well-publicized and extremely well-attended public hearings on issues related to the safety and labeling of food products of biotechnology.[35] In the process, the agency solicited comment from all interested parties. The agency hardly had time to evaluate the comments from those hearings before, in February 2000, 130 nations met in Montreal to forge a global treaty regarding the safety and labeling of genetically manufactured crops (Downs 2000: 3). This meeting resulted in the drafting of a proposed treaty that, among other things, would address the labeling of seeds and raw agricultural commodities that contain genetically modified organisms. Under the treaty, such foods would have to be accompanied by a disclosure that the commodity "may contain" "living" genetic material. The treaty is not effective until a lengthy ratification process is completed. The developments prompted the FDA to issue an updated guidance with respect to the labeling of food products of biotechnology that reconfirmed the agency's past stated positions on the topic (FDA Center for Food Safety and Applied Nutrition 2001).

The FDA's labeling policy with respect to genetically modified foods has been upheld by the courts. Although recognizing that consumers might find it irksome that the FDA does not read section 201(n) as authorizing labeling requirements solely because of consumer demand, the U.S. District Court for the District of Columbia has emphasized that these consumers "fail to understand the limitation on the FDA's power to consider consumer demand when making labeling decisions because they fail to recognize that the determination that a product differs materially from the type of product it purports to be is a factual predicate to the requirement of labeling."[36] Thus, the court observed, if there is a material difference and consumers would likely want to know about the difference, then labeling is appropriate under section 201(n). If, however, the product does not differ in any significant way from what it purports to be, then it would be misbranding to label the product as different even if consumers misperceived the product as different.

The FDA continues to hold the scientific and legal view that information about methods used in the development of a food are neither essential nor material. When asked in 2000 why the agency does not require companies to tell consumers on the food label that a food is the product of biotechnology, the Commissioner of Food and Drugs, Jane Henney, captured the essence of the foregoing discussion in her response:

> Traditional and bioengineered foods are all subject to the same labeling requirements. All labeling for a food product must be truthful and not misleading. If a bioengineered food is significantly different from its conventional counterpart—if the nutritional value changes or it causes allergies—it must be labeled to indicate that difference. For example, genetic modifications in varieties of soybeans and canola changed the fatty acid composition in the oils of those plants. Foods using those oils must be labeled, including using a new standard name that indicates the bioengineered oil's difference from conventional soy and canola oils. If a food had a new allergy-causing protein introduced into it, the label would have to state that it contained the allergen.

> We are not aware of any information that foods developed through genetic engineering differ as a class in quality, safety, or any other attribute from foods developed through conventional means. That's why there has been no requirement to add a special label saying that they are bioengineered. Companies are free to include in the labeling of a bioengineered product any statement as long as the labeling is truthful and not misleading. Obviously, a label that implies that a food is better than another because it was, or was not, bioengineered, would be misleading. (quoted in Thompson 2000)

Even more recently, the FDA echoed the commissioner's observations when the agency took the rare step to formally object to a state ballot initiative (ultimately rejected by the voters) designed to require the mandatory labeling of foods and food additives produced using genetic engineering. The agency advised the Governor of Oregon as follows:

> After numerous meetings and public comments on this issue, FDA concluded that a safety assessment of any new food should focus on the traits and characteristics of that food, no matter which techniques (traditional breeding or genetic engineering) were used to develop the food. Food produced via bioengineering should be treated just like its conventional counterparts because, from a scientific standpoint, there is no evidence that these foods differ as a class from traditionally bred foods in any meaningful or uniform way. Nor is there evidence that, as a class, foods developed by rDNA breeding techniques present any different or greater safety concerns than foods developed via traditional breeding. FDA's scientific evaluation to date has shown that the substances added to food via bioengineering have been well-characterized proteins that are functionally very similar to other proteins that are commonly and safely consumed in the diet every day.
>
> FDA has previously concluded that requiring mandatory labeling for bioengineered foods is not scientifically or legally warranted. Rather, the labeling for foods produced using bioengineering must comply with the law applying to the labeling for all foods. Among other things, food labeling must reveal all facts that are material in light of representation made in the labeling or in light of consequences that may result from the use of foods. 21 U.S.C. § 321(n). (Crawford 2002)

The FDA went on to explain that it would consider mandatory labeling under the following circumstances:

- If the food is significantly different from its traditional counterpart, such that the common or usual name no longer adequately describes the new food—the FDA has required labeling for two foods (a soy oil and a canola oil) where the fatty acid composition was changed to mimic that of food oils not associated with the modified plant
- When an issue exists for the food or a constituent of the food regarding how the food is used or consequences of its use
- If the food has significantly different nutritional properties
- If a new food includes an allergen that consumers would not expect to be present in the food based on the food's name (Crawford 2002)

These formal positions are consistent with prevailing statutory authority and with the FDA's longstanding policy that it should categorically require only clearly essential or material information to appear on the food label.

### Voluntary Labeling

The FDA's concern about collateral information extends even to information that cannot be required, that is, to voluntarily labeled information. Consider, for example, in the case of labeling of milk derived from cows that have been treated with recombinant bovine somatotropin (rBST). There the agency provided "interim guidance" in 1994 that cautioned that even truthful information could mislead consumers[37] and largely reaffirmed that position with its 2001 guidance on voluntary labeling.[38] Accordingly, the agency's guidances emphasize that if voluntary labeling is to be employed, misleading implications must be avoided and the information presented must appear in its proper context. Thus, it is the FDA's view that, under federal law, voluntary representations with regard to the presence or absence of genetic modification in food are potentially misleading. The agency has, correspondingly, advised that any such message must be crafted with care and caution.[39]

There is a precatory element to the FDA's voluntary labeling guidance. Since 1994, a string of court cases involving the doctrine of "commercial speech" has highlighted the agency's responsibility not to unnecessarily restrain "speech."[40] These cases make clear that the FDA has a substantial interest in restricting the dissemination of information that can be proven to have the potential to mislead consumers. But, significantly, the cases clarify that the requisite fit between the government restriction and achieving its substantial interest is not necessarily perfect and must be reasonable. As a result, the FDA must employ not necessarily the least restrictive means in evaluating the propriety of a representation in "voluntary" labeling but a means narrowly tailored to achieve the desired objective. Thus, where the FDA seeks to prohibit a potentially deceptive representation, case law today appears to require the agency to show that the claim actually has the potential to mislead *and* that this potential cannot reasonably be cured by a disclaimer.[41] In reality, this added burden places an enormous strain on the FDA's resources, ability, and desire to attempt to enforce, on a case-by-case basis, adherence to its collective guidance on labeling for products of biotechnology.[42] Simply stated, there is, today, a disconnect between how the FDA would like voluntary labeling to describe products of biotechnology and what the FDA is reasonably able to prohibit purveyors from saying on food labels about products of biotechnology. As a result, the ability of producers to develop and employ what they believe to be nonmisleading "voluntary" labeling with respect to whether a product has been produced via biotechnology is enhanced and, in many cases, for practical purposes, largely unfettered. "False" (as opposed to potentially "misleading") representations, however, with respect to biotechnology remain vulnerable to agency enforcement action.

## DISCUSSION

Responsible debate and discussion can go a long way to address the bases for the desire of some for information about the genetically engineered status of food and to explain the FDA's reluctance to depart from its traditional interpretation of its food labeling authority to *require* such information. To the latter end, it is critical to remember that when the agency does deem information regarding genetically engineered foods to be material—the scenario of a peanut gene in a tomato, for example—its policy is to clearly require that information to appear on the food label. When, however, the only difference involves whether new technologies were used in the manufacture or production of food *and* no meaningful difference has been found to exist between the modified food and its traditional counterpart, the agency has concluded that neither an empirical nor a legal justification exists to depart from its rather clear course of almost 70 years of regulation of the food label since 1938. A further incentive against such a departure may well be the understandable federal desire to reasonably accommodate what it believes to be a safe innovation in food technology in light of the growing awareness about the value biotechnology may have in helping to assure a wholesome and abundant food supply both here in the United States and, perhaps even more important, in less developed countries throughout the world.

For many, the position the FDA has taken with regard to the labeling of food products of biotechnology will continue to grate. How, some ask, can the FDA not provide information consumers wish to know? The answer is simple: The FDA has reasoned that, for the most part, information concerning biotechnology-derived status of a food or food ingredient is not material. This view reflects the agency's scientific judgment as to the safety of the technology and the lack of materiality in difference between a modified food and its traditional counterpart. The corresponding response for many consumers and interested parties is also simple: More assurance of safety is needed, or simpler still, "I just would like to know."

Clearly, efforts by the FDA and the scientific community to frame and communicate their view that there is nothing inherently mysterious or unsafe about genetic engineering would be helpful. Industry, too, can help increase public confidence not only by openly addressing the public's doubts and concerns but also by developing products that deliver real, easily recognizable, and desirable tangible benefits to consumers. But these boil down to issues of public trust, not labeling policy, and, as such, should be kept distinct from the legal issues of essentiality and materiality that have characterized the FDA's implementation of its authority to require information to appear on the food label.

In implementing its labeling policies, the FDA has opted to make case-by-case decisions as to what information is material. Without question, trivial or insignificant facts about a product have not been deemed material. And, something more than a desire to know has been needed to establish an empirical basis for why information is material. In the context of food biotechnology, it thus seems that to establish materiality, at least two questions must be answered: (1) Why is information related to genetic engineering as material as, say, information on gluten where failure to provide could fundamentally mislead a consumer about the basic attributes of a food? (2) What is so different about a genetically engineered food that will render the omission of information about it materially misleading to the consumer?

An alternative, of course, to FDA-required labeling is voluntary labeling by manufacturers. As the opening discussion in this article reveals, much of the information that appears on the food label is voluntarily placed there. The FDA obviously has an interest in prohibiting any voluntary labeling program that is misleading. That function, however, is distinct from exercising section 201(n) to *require* information to appear on the food label: Most of the information voluntarily appearing on the label is well beyond the bounds of essentiality and materiality that would allow the FDA to require it there in the first place.

Although one reason for voluntarily placing information on the label is clearly promotional, another is the manufacturer's willingness to respond to a common desire for certain types of information. It is here, in voluntary labeling, that nonmaterial but nevertheless desired nonmisleading information has traditionally been delivered on the food label. Other information-communication vehicles beyond the food label, for example, the Internet and the use of 1–800 telephone numbers, also exist to respond to consumer interests and concerns. Simply put, appropriate and workable legal pathways for information concerning such interests and concerns are available—pathways that do not interfere with the FDA's longstanding practice and statutory interpretation limiting the nature and amount of information that can, and should, be *required* to appear on the food label.

## Notes

Thanks to the Food and Drug Law Institute for permission to publish a revised version of my article "Biotechnology and the Food Label" (*Food and Drug Law Journal* 55 [2000]).

1. Pub. L. No. 75-717, Section 403(a), 52 Stat. 1047 (codified at 21 U.S.C. § 343, 1938).
2. Pub. L. No. 59-384, 34 Stat. 768 (1906).
3. S. Rep. No. 361, 74th Cong., 1st Sess. (1935); reprinted in Dunn (1938).
4. See 21 U.S.C. §§ 343(e), (g), (i) (Food, Drug, and Cosmetic Act §§ 403(e), (g), (i), 1938).
5. See § 403(k).
6. See, e.g., Dunn (1938: 1224).
7. Pub. L. No. 101-535, § 2, 104 Stat. 2353 (codified at 21 U.S.C. § 343(q)(1), Food, Drug, and Cosmetic Act § 403(q)(ii), 1990).
8. H. Rep. No. 101-538, 101st Cong., 2nd Sess. 18 (1990).
9. 55 Fed. Reg. 29487, 29492 (July 19, 1990).
10. 58 Fed. Reg. 2079, 2107 (January 6, 1993).
11. 58 Fed. Reg. 2079, 2107 (January 6, 1993).
12. 58 Fed. Reg. 2510 (January 6, 1993).
13. 63 Fed. Reg. 17008–17010 (April 4, 2005) and 63 Fed. Reg. 17010–17014 (April 4, 2005).
14. 21 U.S.C. § 321(n) (Food, Drug, and Cosmetic Act § 201(n), 1938). Section 201(n) provides as follows:

> If an article is alleged to be misbranded because the labeling or advertising is misleading, then in determining whether the labeling or advertising is misleading there shall be taken into account (among other things) not only representations made or suggested by statement, word, design, device, or any combination thereof, but also the extent to which the labeling or advertising fails to reveal facts material in the light of such representations or material with respect to consequences which may result from the use of the article to which the labeling or advertising relates under the conditions of use prescribed in the labeling or advertising thereof or under such conditions of use as are customary or usual.

15. 38 Fed. Reg. 2125 (January 19, 1973).
16. 21 C.F.R. § 184.1321 (2006).

17. 21 C.F.R. § 101.17(d) (2006).
18. 55 Fed. Reg. 29847 (July 19, 1990).
19. 56 Fed. Reg. 28592, 28596 (June 21, 1991).
20. 56 Fed. Reg. 28603 (June 21, 1991).
21. 21 C.F.R. §§ 310.501 (oral contraceptives), 801.427 (I.U.D.s) (2006).
22. 21 C.F.R. § 801.430 (2006).
23. 45 Fed. Reg. 69840 (October 21, 1980).
24. 21 C.F.R. § 179.26(c) (2006).
25. 51 Fed. Reg. 13376 (April 18, 1986).
26. 21 C.F.R. § 172.867(e); Final Rule, "Olestra," 61 Fed. Reg. 3117 (January 30, 1996).
27. 61 Fed. Reg. 3159 (January 30, 1996).
28. 61 Fed. Reg. 3160–3162.
29. Notice of Filing of Food Additive Petition. 43 Fed. Reg. 11,585 (March 3, 2000).
30. 68 Fed. Reg. 64,363 (August 5, 2003).
31. 57 Fed. Reg. 22984 (May 29, 1992).
32. 57 Fed. Reg. 22991 (May 29, 1992).
33. 57 Fed. Reg. 22991 (May 29, 1992).
34. European Parliament and Council Directive 258/97, concerning novel foods and novel food ingredients, O.J. (L 43) (1997).
35. 64 Fed. Reg. 57470 (October 25, 1999).
36. *Alliance for Biointegrity v. Shalala*, 116 F. Supp. 166, 179 (D.D.C. September 29, 2000).
37. 59 Fed. Reg. 6279 (February 10, 1994).
38. Draft Guidance: Voluntary Labeling Indicating Whether Foods Have or Have Not Been Developed Using Bioengineering, January 2001 available at www.cfsan.fda.gov/~dms/biolabgu.html.
39. 64 Fed. Reg. 57470 (October 25, 1999).
40. See *Pearson v. Shalala*, 164 F.3rd 650 (D.C. Cir. 1999); *Washington Legal Foundation v. FDA*, 13 F. Supp.2d 51 (D.C. 1999); and *Thompson v. Western States Medical Center*, 535 U.S. 357 (2002).
41. See *Pearson v. Shalala*, supra n. 40.
42. The commercial speech line of cases may well also affect the burden the FDA will face in future efforts based on sections 201(n) and 403(n) to require information to appear on the food label.

## References


Crawford, Lester M. October 4, 2002. "Letter from Lester M. Crawford, Deputy Commissioner, to Governor John A. Kitzhaber," available at www.bio.org/local/foodag/Kitzhaber.pdf

Downs, Terry. February 2, 2000. "International Accord Reached on Trade, Labeling of Living GMOs." *World Food Chemical News*, 6.

Dunn, C.W. 1938. *Federal Food, Drug, and Cosmetic Act, a Statement of Its Legislative Record.* New York: G.E. Stechert and Company.

Hutt, Berton Peter, and Merrill, Richard A. 1991. *Food and Drug Law: Cases and Materials*, 2nd ed. Westbury, NY: The Foundation Press, Inc.

Thompson, Larry. January-February 2000. "Are Bioengineered Foods Safe?" *FDA Consumer Magazine*. Available at www.fda.gov/fdac/features/2000/100_bio.html

U. S. Food and Drug Adminstration Center for Food Safety and Applied Nutrition. January 2001. *Guidance for Industry: Voluntary Labeling Indicating Whether Foods Have or Have Not Been Developed Using Bioengineering. Draft Guidance*. Available at www.cfsan.fda.gov/~dms/biolabgu.html.

# 4

# European Community Legislation for Traceability and Labeling of Genetically Modified Crops, Food, and Feed

*Margaret Rosso Grossman*

In the last several years, European Union (EU) policy has encouraged research in biotechnology, including bioengineered or genetically modified (GM) agricultural crops, and its strategy for life sciences and biotechnology aims to improve the vitality and competitiveness of the European biotechnology sector (COM(2004)250). EU documents acknowledge the significance of GM crops, and a recent report noted that "the potential of plant genomics and biotechnology to deliver major advances in our lifestyles and prosperity is enormous. [Biotechnology] can also maintain and enhance the competitiveness of EU farmers and food producers" (European Commission 2004b: 19).

In practice, however, producers in the European Union have been reluctant to grow GM crops. In 2003, for example, only about 32,000 hectares of GM crops were sown, mostly in Spain (Directorate-General for Agriculture and Rural Development 2005: 4.23); in 2004, the total increased to about 58,000 hectares (Directorate-General for Agriculture and Rural Development 2006: 4.23). In 2006, six EU Member States planted GM crops, for a total of about 68,500 hectares (60,000 in Spain) (James, 2006). Few GM products are sold to European consumers, who prefer not to eat GM foods. Scientists disagree about the risks and benefits of GM crops and food products. European Community legislators, too, disagree and have enacted new regulatory measures only after long deliberation, even as some Member States ban GM crops and foods in their territories.

Nonetheless, after a lengthy moratorium, the European Community (E.C. or Community)[1] began to authorize new GM products. In May 2004, the European Commission authorized marketing of a GM sweet maize, Syngenta's Bt11, for food and food ingredients (Commission Decision 2004/657). This maize had been authorized for feed, processing, and import in 1998. Other products followed, and by early November 2005, four products had been approved under stringent new regulatory measures, but without clear agreement from Member States. For each product, a scientific risk assessment ensured that it poses no danger to the environment, and an assessment by the European Food Safety Authority (EFSA) concluded that it is as safe as a non-GM product (e.g., EFSA 2003). Each must be labeled clearly as "genetically modified," and unique identifiers will ensure traceability through the process of postmarket monitoring. Food and feed products will be entered in the Community Register of Genetically Modified Food and Feed.

In July 2004, after a regulatory process of almost four years, the European Commission approved for 10 years the import and processing of Monsanto's GM maize, NK 603, for use in animal feed and for industrial purposes (Commission Decision 2004/643). The Commission later authorized NK 603 for food and food ingredients (Commission Decision 2005/448), but cultivation is not permitted. In August 2005, the Commission approved two more GM varieties for use in animal feed, but not cultivation or food: Monsanto's maize MON 863, a Bt maize modified for resistance for corn rootworm (Commission Decision 2005/608), and Monsanto's oilseed rape GT 73 (Commission Decision 2005/635).[2] For the latter, the Commission prescribed measures to protect health and the environment in case of accidental spillage during transport, storage, handling, or processing (Commission Recommendation 2005/637). In November 2005, the Commission authorized Pioneer/Mycogen Seeds' maize DAS1507, modified for resistance to lepidopteran pests and glufosinate-ammonium herbicide (Commission Decision 2005/772). In January 2006, Monsanto's maize varieties MON 863 and GA 21 (tolerant to the herbicide glyphosate) were approved for food use, followed by DAS1507 in March 2006. A year later, in March 2007, the Commission authorized three oilseed rape varieties (modified for tolerance to the herbicide glufosinate-ammonium) for import and for processing as animal feed or industrial products, but not for cultivation or food (European Commission 2007). With these measures, at least 20 GM foods and 12 GM feeds had been approved for sale in the European Union. A number of other products await authorization.

Approvals of GM crops for cultivation in the European Union have progressed more slowly. Indeed, no new product has been approved for cultivation in the European Union since 2001 (EuropaBio 2007: 6). For the first time, in September 2004, the Commission listed GM seeds in the Common Catalogue of Varieties of Agricultural Plant Species (see Council Directive 2002/53). The 17 varieties were derived from Monsanto's MON 810 maize, authorized in 1998. Listing in the Common Catalogue allows seeds to be sold and planted in all Member States (European Commission 2004d), but several States have banned their cultivation.

The decisions in 2004 signaled the end of a de facto moratorium since 1998 on approvals of GM crops for import. To some extent, this moratorium had resulted from a perception that regulatory measures were inadequate to govern GM crops, food, and feed. In 2001, after a lengthy legislative process, the European Community began to enact new measures to regulate genetically modified organisms (GMOs) when they are placed on the market and afterward, and to ensure that GMOs and GM products can be identified and traced through their life cycle with labels to provide adequate information and consumer choice.

The most important measures directed specifically toward GMOs are as follows:

- Directive 90/219 on the contained use of GM microorganisms (GMMs)
- Directive 2001/18 on the deliberate release into the environment of GMOs and repealing Council Directive 90/220
- Regulation 1829/2003 on GM food and feed
- Regulation 1830/2003 concerning the traceability and labeling of GMOs and the traceability of food and feed products produced from GMOs

To complete the system, the Commission proposed a decision to establish thresholds for traces of GM seeds in other seed products (Draft Commission Decision), but the measure has not been agreed.

Other E.C. measures affect GMOs, too. For example, the General Food Law (Regulation 178/2002) sets out general principles and establishes the EFSA. In addition, the Environmental Liability Directive allocates responsibility for some types of environmental damage from GMOs (Directive 2004/35).

## LAWMAKING IN THE EUROPEAN COMMUNITY

### Legislation

The European Union, 27 Member States as of January 2007, is governed by primary legislation—its founding Treaties, as amended—and by secondary legislation. The Treaty of Rome (1958, with later amendments) governs the European Community. It establishes Community institutions, authorizes Community activities (e.g., agriculture, environment), and prescribes procedures for decision making.[3]

The European Community enacts secondary legislation under authority of the Treaty. Measures that govern GMOs invoke Treaty authority to regulate in the areas of agriculture, the environment, and public health (E.C. Treaty, arts. 37, 175, 172), as well as authority for the harmonization of Member State laws that affect the internal market (art. 95).

Genetically modified organisms are governed by several different types of secondary legislation. Most important are regulations and directives. Regulations have general application; they are binding in their entirety and apply directly in all Member States (E.C. Treaty, art. 249). Most regulations need not be transposed into State law, but regulations that govern GMOs require cooperation of States in the authorization process. Directives are binding as to the result to be achieved but leave the choice of form and methods to Member States (art. 249). Directives are normally effective after implementation in State law. The European Commission may seek enforcement against Member States that fail to enact timely implementing provisions (art. 226); in a case against France, for example, the Court of Justice held that France had failed to transpose the 1990 Directive that governed deliberate release of GMOs (*Commission v. French Republic* 2003).

Other measures include decisions, recommendations, and opinions. Decisions bind the Member States or private parties to whom they are addressed, while recommendations and opinions have no binding force (E.C. Treaty, art. 249).

The Treaty prescribes several methods for enacting directives and regulations. Recent measures that govern GMOs have been enacted under the co-decision procedure (art. 251), which requires agreement of both the European Parliament and the Council of the European Union. For most measures, the Council must agree by a qualified majority, a system of weighted Member State votes prescribed by Treaty. In addition, the Commission or the Council may issue detailed rules to implement measures enacted by Parliament and Council.

The European Court of Justice and Court of First Instance (E.C. Treaty, arts. 220–245) ensure that "in the interpretation and application of this Treaty the law is observed" (art. 220). The Treaty establishes the jurisdiction of each court. The Commission, national courts, and individuals may refer matters to the courts.

## Precaution in GM Legislation

Part of European concern about GMOs is environmental. Though the Treaty of Rome did not provide E.C. competence for environmental matters, the Single European Act (1987) added a title on the environment and provided a clear legal basis for enacting environmental measures (for details, see Grossman 1995). The environmental title of the Treaty indicates that E.C. environmental policy should "aim at a high level of protection" and should be based on "the precautionary principle and on the principles that preventive action should be taken, that environmental damage should as a priority be rectified at source and that the polluter should pay" (E.C. Treaty art. 174(2)). The integration principle prescribes that environmental protection be integrated into the definition and implementation of other E.C. policies and activities (art. 6). These principles, especially the precautionary principle, influence E.C. regulation of GMOs.

The precautionary principle (or precautionary approach) has become part of international law, particularly in measures that protect the environment. For example, the Rio Declaration (prin. 15) invokes the principle, and both the Convention on Biological Diversity (preamble) and the related Cartagena Protocol on Biosafety (art. 11(8)) advocate the precautionary approach (see Bergkamp 2003: 157–214; de Sadeleer 2002: 91–223). The principle also applies in other fields, including food law and public health. It is controversial, and it is not uniformly accepted, interpreted, or applied.

In the European Community, the Treaty adopts, but does not explain, the precautionary principle. The *Communication from the Commission on the precautionary principle* elaborates:

> Although the precautionary principle is not explicitly mentioned in the Treaty except in the environmental field, its scope is far wider and covers those specific circumstances where scientific evidence is insufficient, inconclusive or uncertain and there are indications through preliminary objective scientific evaluation that there are reasonable grounds for concern that the potentially dangerous effects on the environment, human, animal or plant health may be inconsistent with the chosen level of protection. (COM(2000)1: 9–10)

The principle, used to manage risk, is triggered by unacceptable risk and scientific uncertainty. Because biotechnology is perceived to pose uncertain risks to health and the environment, GMOs invite application of the precautionary principle, and much commentary has focused on its application to the management of risk from GMOs (Forsman 2004).

Early E.C. legislation that governed GMOs used a precautionary approach, and precaution was one basis for the de facto moratorium on authorizations of GM varieties (see Francescon 2001: 311). Under the new regulatory system, Directive 2001/18 invokes the precautionary principle explicitly in the preamble and indicates that Member States "shall, in accordance with the precautionary principle, ensure that all appropriate measures are taken to avoid adverse effects on human health and the environment which might arise from the deliberate release or the placing on the market of GMOs" (art. 4(1)). Various provisions of Directive 2001/18, especially for risk assessment and postapproval monitoring, implement the principle.

Regulation 1830/2003 on traceability and labeling, imposes "risk management measures in accordance with the precautionary principle" (preamble (3)). The accompanying GM food and feed Regulation 1829/2003 refers instead to a "high level of protection of human life and health" (preamble (2)).

The 2002 General Food Law, too, devotes an article to the precautionary principle. When "the possibility of harmful effects on health is identified but scientific uncertainty persists, provisional risk management measures necessary to ensure the high level of health protection chosen in the Community may be adopted, pending further scientific information for a more comprehensive risk assessment" (Regulation 178/2002: art. 7(1)). Measures should be proportionate, and "no more restrictive of trade than is required … , regard being had to technical and economic feasibility and other factors" (art. 7(2)). The regulatory system for GMOs established by these measures ensures precaution.

## E.C. AUTHORIZATION OF GM PRODUCTS

European Community regulation of GMOs and GM food and feed relies on a number of interrelated directives and regulations enacted and amended since 1990. Authorization of GMOs is process based, rather than product based. It requires case-by-case evaluation of GMOs and follows a step-by-step process of decreasing containment. Traceability and labeling, with thresholds for applicability, are critical components of recent E.C. regulation.

### Contained Use: Council Directive 90/219

Council Directive 90/219, which continues in force, governs contained use of genetically modified micro-organisms (GMMs). Contained use refers to activities in which microorganisms are genetically modified or in which GMMs are used and for which "specific containment measures are used to limit their contact with, and to provide a high level of safety for, the general population and the environment" (art. 2(c)). The Directive governs research, work in laboratories, and industrial work with GMMs.

Member States designate a competent authority to implement the Directive. To avoid adverse effects on health and environment from contained uses of GMMs, users carry out risk assessments to determine the class of risk (negligible, low, moderate, or high) and the resulting assignment of containment level (arts. 5, 6). Moderate- or high-risk contained use of GMMs requires prior written consent of the competent authority (art. 10). Users apply containment and other protective measures for each class of contained use and notify Member State authorities of activities on their premises (arts. 7–12). Member States notify the Commission and other States when accidents occur and report to the Commission annually (with a summary every three years).

Annexes to the Directive describe criteria to be met when establishing safety of GMMs for human health and the environment; principles to be applied in the risk assessment; containment measures for laboratories, growing rooms, animal units, and other activities; and content of notifications under the Directive (Annexes II–V). (For more detail on Council Directives 90/219 and 90/220, see Grossman and Endres 2000: 393–403.)

### Deliberate Release

The European Community enacted Directive 2001/18 on deliberate release after a decade of experience with its predecessor, Council Directive 90/220. The new Directive was intended to make the authorization of GMOs more efficient and transparent and to control risks that might threaten human health and the environment (preamble (5), (48); see Brosset 2004). It imposed more stringent measures for environmental risk assessment, added a postmarket monitoring requirement, limited authorization for release of GMOs to 10 years, and required Member States to ensure traceability and labeling of GMOs at all stages of placing on the market. Other regulatory measures provide detailed guidance notes (Council Decision 2002/811, Commission Decision 2002/623), formats for submitting information (Council Decisions 2002/812, 2002/813; Commission Decision 2003/701), and requirements for GMO registers (e.g., Commission Decision 2004/204).

Member States were to have implemented Directive 2001/18 in their national laws by October 17, 2002, but not all States have done so. By August 2004, 7 of the 15 Member States and 8 of the 10 new States had communicated implementation measures. The Commission filed legal actions against 8 of the 15 States for failure to enact national measures (COM(2004)575: 4). By October 2006, the EUR-Lex website listed implementing measures in 22 States (but without indicating whether the measures are comprehensive or in conformity with the Directive).

Under the Directive, a GMO is "an organism, with the exception of human beings, in which the genetic material has been altered in a way that does not occur naturally by mating and/or natural recombination" (art. 2(2)). Anyone who plans to seek authorization for release of a GMO must carry out an environmental risk assessment (art. 6). The Directive identifies types of information that might be needed to carry out the risk assessment for higher plants and other organisms (Annex III). Potential adverse effects of the release on human health and the environment, especially from gene transfer, must be assessed under "principles" for risk assessment (Annex II). More specifically, applicants should evaluate possible adverse effects, their likelihood, and the magnitude of the consequences, and identify management strategies to reduce risks (Commission Decision 2002/623).

Directive 2001/18 governs deliberate release of GMOs in two circumstances: "any other purpose than for placing on the market" and "placing on the market of GMOs as or in products." Regulation 1829/2003 governs GM food and feed, and the "one door–one key principle" allows a single authorization for deliberate release of a GMO and its use as food or feed.

## Nonmarket Deliberate Releases

Directive 2001/18 implements a step-by-step principle. That is, "the containment of GMOs is reduced and the scale of release increased gradually, step by step, but only if evaluation of the earlier steps in terms of protection of human health and the environment indicates that the next step can be taken" (preamble (24)). Genetically modified organisms cannot be placed on the market until they have been field tested in appropriate ecosystems; therefore, deliberate release for research is a step toward marketing (preamble (23), 25).

Member States, through their competent authorities, authorize releases for field testing or other research. The Directive requires notification to the Member State competent authority before deliberate release of a GMO or combination of GMOs. Notification includes a detailed technical dossier (Annex III; see also Council Decision 2002/813: art. 1) and an environmental risk assessment (Directive 2001/18: art. 6). A summary of each notification is sent to the Commission, which forwards the summary to other Member States, who may "present observations" (art. 11). If appropriate, Member States may consult the public on a proposed release (art. 9(1)). The State competent authority gives written consent to the release or, if the proposed release does not comply with the Directive, rejects the notification.

No release is permitted until the notifier has received written consent from the competent authority. After release, the notifier must report results, especially risks to health or the environment, to the competent authority (art. 10; Commission Decision 2003/701). Member States provide public information on all releases in their territory, but without disclosing confidential information (art. 9(2)).

## Placing GM Products on the Market

Placing products on the market affects the entire European Community. Once written consent for a GMO has been issued, that GMO may be used "without further notification throughout the community" (Directive 2001/18: art. 19(1)). Member States may not "prohibit, restrict or impede the placing on the market of

GMOs, as or in products," if the GMOs have been authorized (art. 22), though a safeguard clause allows State restrictions in limited instances (art. 23). Therefore, procedures for placing GMOs on the market are more complicated and require more involvement of the Commission and Member States. Similar provisions govern GM food and feed under Regulation 1829/2003.

Moreover, these measures apply to imports, and products containing or consisting of GMOs cannot be imported into the European Community if they do not comply with E.C. requirements (Directive 2001/18: preamble (11); Regulation 1829/2003: preamble (43)).

## GMOs: Directive 2001/18

The authorization procedure for marketing GMOs begins with notification to the Member State competent authority, providing descriptions of the GMO, the environmental risk assessment, a plan for postrelease monitoring, conditions for use and handling, and other information (art. 13(2)). A summary of the dossier must indicate the nature of the GMO and its predicted behavior, plus information about previous releases and the monitoring plan (Council Decision 2002/812: Annex).

The postrelease monitoring plan is an important component of notification (Directive 2001/18: Annex VII). Its objectives are to ensure that assumptions underlying the environment risk assessment are correct and to identify unanticipated adverse effects on human health or the environment. The unique characteristics of each GMO mean that monitoring plans must be developed on a case-by-case basis (Council Decision 2002/811: Annex).

The Member State authority examines each notification for compliance with the Directive and then prepares an assessment report, which focuses particularly on the proposed use, the environmental risk assessment, and the proposed postrelease monitoring plan (Directive 2001/18: Annex VI). A competent authority may decide that a GMO should not be placed on the market and reject the notification (art. 15). If the authority concludes that the GMO should be placed on the market, it sends the dossier summary, along with its assessment report, to the Commission and to the competent authorities of other Member States (art. 4). The Commission makes the dossier summary and assessment report available for public comment (art. 24(1)) and prepares registers of genetic modifications. To protect commercial interests, confidential data are accessible only to Member States, the Commission, and the EFSA, but not to the public (Commission Decision 2004/204).

The Commission and other States may ask for information, make comments, or "present reasoned objections" to the placing of a GMO on the market (art. 15(1)). If no objections are made, or if outstanding issues are resolved, the competent authority that assessed the GMO may give written consent to the notifier and also inform the Commission and the other Member States (art. 15(3)). Indeed, in a case brought under Council Directive 90/220, the European Court of Justice held that when no objections are raised, the competent authority is obliged to give consent (*Association Greenpeace France v. Ministère de l'Agriculture et de la Pêche* 2000; see Francescon 2001: 312).

Written consent will include specific conditions for use, handling, and packaging of the GMO or protection of the environment. It will require labeling ("This product contains genetically modified organisms") and monitoring (art. 19(3)). After consent, notifiers follow the prescribed monitoring plan and report regularly to the Commission and competent authorities; results of monitoring are also available to the public (art. 20). Authorization of a GMO lasts 10 years at most and may be renewed (arts. 15(4), 17).

The procedure just described applies unless the Commission or a Member State raises and maintains an objection. Indeed, in most cases at least one State has objected (see Brosset 2004: 568–71). Objection triggers a more complicated procedure, and the Commission then consults the competent scientific committee, EFSA's Scientific Panel on Genetically Modified Organisms (art. 28). If the scientific decision is favorable, the Commission will follow the Community interagency regulatory procedure (the so-called comitology procedure; Council Decision 1999/468) to reach a decision (arts. 18, 30(2)). The Commission submits a draft of the measure to be taken (i.e., its legislative decision to authorize the proposed GMO) to a regulatory committee, made up of Member State representatives. If that committee agrees, the Commission will grant consent. If not, the Commission submits the measure to the Council for decision (and informs Parliament). If the Council does not either agree or oppose the consent by a qualified majority, the Commission may grant consent. The Commission used this procedure for recent authorizations of GM products.

Even after consent is granted, a safeguard clause protects Member States (art. 23). A State may restrict or prohibit use or sale of a GMO if it has grounds to conclude that the GMO poses a risk to human health or the environment, on the basis of either information made available since the date of consent or a reassessment of existing data using new scientific information. The State informs the Commission and other States, and the Commission, with assistance of the Scientific Committee, decides whether the State's action is justified. Under Council Directive 90/220 (art. 16), six Member States invoked the safeguard clause in nine attempted bans of GMOs (European Commission 2005a). In each instance, no justification for the State ban was found. In July 2004, for example, EFSA published opinions of the Scientific Panel on Genetically Modified Organisms that found no new scientific evidence to justify prohibition of certain GM crops in Greece or Austria (EFSA 2004a, 2004b).

The Treaty offers an additional safeguard: Member States may introduce national provisions after adoption of a Council or Commission harmonization measure, "based on new scientific evidence relating to the protection of the environment ... on grounds of a problem specific to that Member State arising after the adoption of the harmonisation measure" (art. 95(5)). The Member State notifies the Commission of the proposed provisions and the grounds for enacting them; the Commission ascertains whether the provisions discriminate or restrict trade and whether they interfere with the internal market (E.C. Treaty art. 95(5),(6)). The Treaty includes an analogous provision for existing Member State measures (art. 95(4)).

Relying on the Treaty safeguard, the Province of Upper Austria proposed to ban cultivation of GMOs to protect traditional and organic production systems, nature, the environment, and biodiversity. To evaluate the scientific justification for the proposed ban, the Commission asked the advice of EFSA, which consulted the Scientific Panel on Genetically Modified Organisms. EFSA concluded that Austria's justification did not meet the requirements of the Treaty. Austria did not provide new scientific evidence related to protection of the environment, prove that its concerns about the coexistence of organic and GM crops were environmental, or show that Upper Austria had unique ecosystems. The Commission therefore rejected Austria's proposed provisions (Commission Decision 2003/653).

Both the Republic of Austria and the Province of Upper Austria sued the Commission, seeking nullification of the decision. The cases were consolidated in the Court of First Instance, which issued its judgment in October 2005 (*Land Oberösterreich v. Commission* 2005). After reviewing the relevant law, the measure at issue, the EFSA opinion, and the Commission decision, the court concluded that the proposed ban on GMOs did not meet Treaty requirements. Upper Austria did not demonstrate the existence of a specific problem that arose after adoption of Directive 2001/18 (*Land Oberösterreich*: ¶¶ 59–69).[4] Both Austria and Upper Austria appealed the decision to the Court of Justice. If affirmed, the decision has implications for the 164 regions and 4,500 municipalities in the European Union that have declared themselves free of GM products (Buenderman 2005).

## Food and Feed: Regulation 1829/2003

Regulation 1829/2003 on GM food and feed protects human and animal health through measures for authorization, supervision, and labeling of GM food and feed (art. 1). For GM foods, it replaces Regulation 258/97 on novel foods, used in the 1990s to authorize several GM foods. The Regulation (preamble (9)) relies on principles articulated in Directive 2001/18 and the risk assessment framework from the 2002 General Food Law (Regulation 178/2002). EFSA administers Regulation 1829/2003 and assesses the risks of GM food or feed, with assistance from Member State agencies (art. 6). The Scientific Panel on Genetically Modified Organisms, established in the General Food Law (art. 28(4)(d)), plays a key role in risk assessment.

Under the one door–one key principle, a single application may cover a GMO and a food or feed containing or consisting of that GMO. To use the one door–one key approach, the applicant provides information normally required under Directive 2001/18—technical dossier, information and conclusions about risk assessment, and a monitoring plan for environmental effects (art. 5(5))—as well as information required by the food and feed Regulation.

The Regulation sets out separate but similar measures for GM food and feed (arts. 3–14, 15–26); products likely to be used both as food and feed may be authorized under a single application (art. 27). In November 2004, EFSA published a lengthy guidance document for preparation of risk assessments of GM plants and GM-derived food and feed (EFSA 2004c).

Regulation 1829/2003 governs GMOs for food use, food containing or consisting of GMOs, and food produced from or containing ingredients produced from GMOs (art. 3(1)). Produced *from* GMOs means "derived, in whole or in part, from GMOs, but not containing or consisting of GMOs" (art. 2(10); see also Regulation 1830/2003: art. 3(2)). The Regulation also governs GM food additives and flavorings (Regulation 1829/2003: preamble (12), (13)).

Products that are produced *with* a GMO but have no GM material in the end product are excluded from regulation. This includes food made with GM processing or products from animals fed with GM feed (preamble (16); Mansour and Key 2004: 61). Genetically modified seeds are governed by other measures and generally fall outside the scope of the Regulation (preamble (34), art. 6(3)(c)). Regulation 1829/2003 applies in a "non-discriminatory manner" to E.C. and imported products (preamble (43)), and it takes account of commitments to international trade and obligations under the Cartagena Protocol on Biosafety (art. 44).

To be placed on the market in the European Union, GM food may not have adverse effects on health or the environment, mislead the consumer, or differ in a nutritionally adverse way from the food it replaces. The authorization process is intended to ensure that these requirements are met (art. 4).

Authorization is similar to the process under Directive 2001/18, though EFSA, rather than the Commission, plays a central role. The applicant submits an application, accompanied by scientific studies, a summary dossier, and other information to the competent authority in a Member State. Commission rules, enacted in consultation with EFSA, guide preparation of applications (Commission Regulation 641/2004). The Member State competent authority sends the application to EFSA, which forwards it to other States and the Commission, making the summary dossier available to the public (Regulation 1829/2003: art. 5).

Authorization requires a scientific evaluation followed by a risk management decision (preamble (9)). EFSA prepares its opinion on the basis of scientific analysis and consultation with experts and (for GMOs, under the one door–one key procedure) with Member State competent authorities. EFSA forwards its opinion, along with an assessment report, to the applicant, the Commission, and the Member States. The opinion is made public, and comments may be submitted to the Commission. The Commission then submits a draft decision on the authorization to the Standing Committee on the Food Chain and Animal Health (arts. 7, 35). The comitology procedure, mentioned above (Council Decision 1999/468), is used to reach final decision on the application. Authorization of GM food is valid throughout the European Union for 10 years and can be renewed (Regulation 1829/2003: arts. 7(5), 11). Authorized GM foods are listed in the Community Register of Genetically Modified Food and Feed (arts. 7(5), 28).

Regulation 1829/2003 also governs GMOs used for feed, feed containing or consisting of GMOs, and feed produced from GMOs (art. 15(1)). Like GM food, GM feed cannot be placed on the market, used, or processed in the European Union without authorization (art. 16(2)). Authorized GM feed must not have adverse effects on human or animal health or on the environment, mislead consumers, impair "distinctive features of the animal products," or differ in a nutritionally adverse way from the feed it replaces (art. 16(1)).

Authorization follows a process similar to GM food, with the initial application, accompanied by scientific studies and other pertinent information, submitted to the Member State competent authority. The authority then informs EFSA, which makes information available to other States, the Commission, and the public (art. 17). The application will undergo stringent scientific evaluation, followed by a risk management decision (preamble (9)). EFSA prepares an opinion, based on scientific evidence and expert consultation, which is sent to the applicant, the Commission, and the Member States; the public may submit comments to the Commission (art. 18(3), (6)–(7)). In light of the EFSA opinion, the Commission submits a draft decision on the application to the Standing Committee on the Food Chain and Animal Health (arts. 19, 35). As with GM food, the Commission uses the comitology procedure to reach a final decision on the application. A GM feed authorization is valid throughout the European Union for 10 years and is renewable, and authorized GM feeds are included in the Community Register of Genetically Modified Food and Feed (arts. 19(5), 28).

## TRACEABILITY AND LABELING: REGULATIONS 1830/2003 AND 1829/2003

Traceability and labeling are critical components of the regulatory system for GMOs in the European Community. Regulations enacted in 2003 work with other measures to require traceability and labeling for GM crops and products.

### Traceability and Labeling in Other Measures

The General Food Law defines "traceability" as "the ability to trace and follow a food, feed, food-producing animal or substance intended to be, or expected to be incorporated into a food or feed, through all stages of production, processing and distribution" (Regulation 178/2002: art. 3(15)).[5] The General Food Law demands a comprehensive system of traceability in the food chain (not only for GMOs) to avoid disruption in case of food safety problems (preamble (28)). It calls for traceability of food, feed, food-producing animals, and other substances used in food and asks food and feed business operators to implement systems and procedures, including labeling, for traceability (art. 18). Guidelines help to implement traceability requirements under the General Food Law (European Commission 2004a). In addition, the European Rapid Alert System for Food and Feed, authorized by the General Food Law (arts. 50–54), aids traceability; for example, in November 2004 the alert system facilitated tracing of dioxin-contaminated potato by-products used for animal feed and blocked movement of animals fed with contaminated feed (European Commission 2004c).

For GMOs specifically, Directive 2001/18 requires traceability "at all stages of the placing on the market of GMOs as or in products authorized under [the placing on the market provisions] of this Directive" (preamble (42)). In addition, both the initial notification procedure and the written consent provisions require compliance with labeling requirements and language that "this product contains genetically modified organisms" (arts. 13(2)(f), 19(3)(e), Annex IV). Member States must ensure that labeling requirements are followed and may establish minimum thresholds for labeling (art. 21).

While Directive 2001/18 calls for a general premarket traceability system for GMOs, it does not define traceability, articulate its objectives, or prescribe an approach for implementation. Moreover, labeling provisions in Directive 2001/18 do not apply to operators who place their GMO products on the open market (MacMaoláin 2003: 874; Francescon 2001: 315).

## Traceability in Regulation 1830/2003

Building on Directive 2001/18, Regulation 1830/2003 applies traceability and labeling "at all stages of the placing on the market" to (nonmedicinal) products consisting of or containing GMOs, food produced from GMOs, and feed produced from GMOs (art. 2(1)). It defines "traceability" (in language slightly different from the General Food Law) as "the ability to trace GMOs and products produced from GMOs at all stages of their placing on the market through the production and distribution chains" (art. 3(3)). The Regulation establishes a unified system of traceability for GMOs and for food and feed products produced from GMOs, with the objectives of "facilitating accurate labelling, monitoring the effects on the environment and, where appropriate, on health, and the implementation of the appropriate risk management measures including, if necessary, withdrawal of products" (art. 1).

Member States ensure compliance through control measures, including sample checks and testing (art. 9(1)–(2)), following the Commission's technical guidelines for sampling and detection of GM material (Commission Recommendation 2004/787).

### Unique Identifiers

Before most provisions of Regulation 1830/2003 could take effect, the Commission had to establish a system of unique identifiers for GMOs. All GMOs placed on the market are to have a unique identifier, registered with the Commission and with the Biosafety Clearing-House established in connection with the Cartagena Protocol on Biosafety (Commission Regulation 65/2004: arts. 2–3; Biosafety Clearing-House). The applicant develops the unique identifier, to be specified in the authorization for the GMO. Genetically modified organisms authorized before January 2004 must also have identifiers.

Formats for unique identifiers are coordinated with the Organisation for Economic Co-operation and Development's BioTrack product database (OECD). Each identifier will have nine alphanumeric digits, divided into three components separated by hyphens. The first component, two or three digits, identifies the applicant or consent holder. The second component, five or six digits, represents the transformation event; a unique number applies to similar transformation events developed by different organizations or in different organisms. The final component, a single verification digit, is calculated from the numerical values of the other digits (Annex). For example, the identifier for Syngenta's Bt11 is SYN-BT-011-1; for Monsanto's NK 603, MON-00603-6; for Pioneer's DAS1507, DAS-01507-1.

### Products Consisting of or Containing GMOs

An "operator" is one who places a product on the market or receives a product placed on the market (art. 3(5)). Operators must ensure that information prescribed for traceability is transmitted in writing, at the first stage of placing on the market and at all subsequent stages. Operators must have systems and standardized procedures to preserve this information and the identity of operators by whom and to whom the products were made available, for five years from each transaction (art. 4A).

Two types of information are required: a statement that the product contains or consists of GMOs, and the unique identifier(s) assigned to the GMOs. For mixtures of GMOs intended for direct use as food or feed or for processing, operators may transmit a declaration of use and a list of unique identifiers (art. 4A). For products with lot numbers or other identification systems prescribed by E.C. legislation, and when lot numbers are clearly marked on the package, certain operators may choose to retain lot number information (art. 6).

### Food and Feed Produced from GMOs

Requirements are less stringent for food and feed produced from GMOs. When placing the product on the market, the operator must transmit in writing each food ingredient produced from GMOs and each feed material or additive produced from GMOs; if no list of ingredients exists, the operator must indicate that the product is produced from GMOs. Unique identifiers are not required. The same five-year retention period applies for information transmitted and identity of operators, with the same lot-number alternative (arts. 5, 6).

## Labeling: Regulations 1830/2003 and 1829/2003

Labeling, which facilitates traceability, applies at all stages of placing on the market. Specific rules for GMOs and GM products do not affect requirements in other E.C. legislation (e.g., Council Directive 2000/13 on labeling and advertising), which continue to apply (Regulation 1830/2003: art. 4A(5); Regulation 1829/2003: art. 13(1)).

### Food

Regulation 1830/2003 requires labels for products consisting of or containing GMOs (art. 4B). Operators must use the words "This product contains genetically modified organisms" or "This product contains genetically modified [name of the organism(s)]." For prepackaged products, those words must appear on the label; for bulk products, on or in connection with the product display.

Regulation 1830/2003 does not govern labels for food or feed produced from GMOs. Instead, Regulation 1829/2003 requires labels for GM foods and feeds delivered to final consumers or mass caterers that either contain or consist of GMOs or are produced from or contain ingredients produced from GMOs (arts. 12–13, 24–26). Labeling applies to GM foods, even when the GM component cannot be detected (Mansour and Key 2004: 62).

Regulation 1829/2003 is more detailed than Regulation 1830/2003 about language for the label. If food consists of more than one ingredient, the words "genetically modified" or "produced from genetically modified [name of ingredient]" must appear in parentheses following the ingredient concerned. If the ingredient is designated by category, the words "contains genetically modified [name of organism]" or "contains [name of ingredient] produced from genetically modified [name of organism]" must appear in the list of ingredients. Further, if there is no list of ingredients, "genetically modified" or "produced from genetically modified [name of organism]" must appear clearly on the label. For bulk food, or pre-packaged food in small containers, GM labels can be on or next to the food display or on the package in a legible font (art. 13(1)).

Labels must mention characteristics or properties of food that differ from its conventional counterpart in composition, nutritional value or effects, intended use, or health implications, or that may raise ethical or religious concerns (art. 13(2)). If food has no conventional counterpart, the label must contain information about its nature and characteristics (art. 13(3)).

### Feed

Labeling rules for feed are similar, though not identical. Rules apply to "[e]ach feed of which a particular feed is composed" (art. 25(2)). Genetically modified organisms for feed use and feed containing or consisting of GMOs must indicate "genetically modified [name of organism]" in parentheses following the name of the feed (or in a footnote to the list of feed). Feed produced from GMOs will indicate "produced from genetically modified [name of organism]." The Regulation does not require labeling of products produced with GMOs or products from animals fed with GM feed, an omission that has triggered criticism (see MacMaoláin 2003).

As with GM food, feed labels must mention any characteristic that is different from the conventional counterpart in composition, nutritional properties, intended use, or health implications, or that may lead to ethical or religious concerns (art. 25(2)(c)).

## Labeling and Traceability Thresholds

A temporary exemption applies to products that contain "adventitious or technically unavoidable" traces of *authorized* GMOs. These may not require labels, and some traceability requirements do not apply. To show that the presence of GM material is adventitious or technically unavoidable, operators must be able to prove that they took appropriate steps to avoid the presence of GM material (Regulation 1830/2003: arts. 12(3), 24(3)).

Under Regulation 1830/2003 (art. 4C(7)), "traces of GMOs in products" do not trigger traceability and labeling requirements if the traces do not exceed the threshold set in Directive 2001/18. Under that Directive (art. 21(3)), if products are intended for direct processing, labels are not required for adventitious or technically unavoidable traces of authorized GMOs of no more than 0.9%.

For traces of GMOs in "products intended for direct use as food, feed or for processing" (art. 4C(8)), the traceability and labeling threshold comes from Regulation 1829/2003. That Regulation, too, establishes a threshold of 0.9% for traces of GMOs in food ingredients considered individually or food consisting of a single ingredient (art. 12(2)),[6] and 0.9% for feed and each feed ingredient (art. 24(2)). Under both measures, lower thresholds can be established.

Regulation 1829/2003 applies a three-year transitional threshold for some *unauthorized* GM material. If authorization is pending and the risk evaluation for a GMO is favorable, the adventitious or technically unavoidable presence of no more than 0.5% of that GM material will not breach the Regulation (art. 47(1)). The Commission maintains a list of GMOs that have received a favorable opinion from the scientific committee (Commission Regulation 641/2004: art. 18(1); European Commission 2006). This transitional exemption, effective only until 2007, also applies to the traceability requirements for food and feed produced from GMOs (Regulation 1830/2003: art. 5(4)). In addition, Directive 2001/18 applies the transitional threshold to exempt certain products—those with traces of GMOs and intended for direct use as food or feed or for processing—from the complex authorization process for placing on the market (art. 12a).

A threshold for authorized GM seeds, required by Directive 2001/18 (art. 21(2)), has not yet been established. Proposed thresholds were 0.3% for cross-pollinating crops (maize, oilseed rape) and 0.5% for self-pollinating species, calculated to allow harvested material to meet the 0.9% threshold for direct use or direct processing (Draft Commission Decision). This proposed measure remains controversial, with the Danish and Austrian delegations demanding a threshold of 0.1%, the detection level for GM seeds (European Council 2004). Both seed industry professionals and organizations that oppose GMOs favor enactment of thresholds for the adventitious presence of GM seeds in conventional seed.

## COEXISTENCE AND LIABILITY

Now that the moratorium on GM crops has ended, with GM varieties added to the Common Catalogue of Varieties of Agricultural Plant Species, producers may legally plant authorized GM crops. But pollen drift from GM crops may result in cross-pollination if farmers sow GM crops in close proximity to conventional or organic crops, and commingling of GM and traditional seeds may occur during production, harvesting, and marketing. Adventitious GM content in conventional or organic crops triggers traceability and labeling requirements. This situation raises the difficult issue of coexistence and the related question of liability for cross-pollination and other damage.

## Coexistence

Coexistence refers to the "admixture" of GM and non-GM crops and to the related "ability of farmers to make a practical choice between conventional, organic and GM-crop production, in compliance with the legal obligations for labelling and/or purity standards" (Commission Recommendation 2003/556: preamble (3)). The probability of admixture of GM and non-GM crops and the measures for avoiding it depend on the type of crop and on geographic factors like natural conditions and field sizes.[7]

The only GM crops that can be cultivated in the European Union have been authorized through a process designed to protect the environment and human health. Therefore, Commission officials insist, coexistence should focus on the economic and legal consequences of adventitious presence of GM crops in non-GM crops, which can result from "seed impurities, cross-pollination, volunteers…harvesting-storage practices and transport" (Fischler 2003: 2; European Commission 2003). Economic effects of adventitious presence of GMOs include, for example, the requirement that conventional crops with GM content be labeled, with a resultant loss of income, or the contamination of organic crops, which cannot be produced from GMOs under E.C. law. Costs of management measures to avoid adventitious presence will affect farmers and seed companies (Fischler 2003: 2–5).

### Commission Recommendation

Directive 2001/18 authorizes Member States to take "appropriate measures to avoid the unintended presence of GMOs in other products." In addition, it directs the Commission to "develop guidelines on the coexistence of genetically modified, conventional and organic crops" (art. 26). The Commission issued those guidelines in its July 2003 Recommendation on coexistence. As a starting point, the Commission stated that "[n]o form of agriculture, be it conventional, organic, or agriculture using [GMOs] should be excluded" in the European Union (Commission Recommendation 2003/556: preamble (1)). Farmers should be free to choose the type of crops they grow, but consumer choice should be protected, too (Annex § 1.1). Because farm structures and natural conditions differ in the various Member States, each State should develop measures for coexistence, with guidance from the Commission (preamble (4)–(8)).

Commission guidelines, which are nonbinding recommendations to Member States, address commercial seed and crop production. They set out general principles to apply and factors to consider in designing State-specific measures. Among the 12 principles are transparency, stakeholder involvement, science-based decision making, a system built on existing means of crop segregation, focus on authorized

GM varieties, and consideration of liability rules (Annex § 2.1). The Commission suggested that "during the phase of introduction of a new production type in a region, operators (farmers) who introduce the new production type should bear the responsibility of implementing the farm management measures necessary to limit gene flow" and should inform neighboring farmers of their plans to plant GM crops (§ 2.1.7).

Factors that States should consider in designing national strategies and prescribing best management practices focus on such practical concerns as geographic level of coexistence (e.g., neighbors or regions), sources of admixture, threshold values for labels, and characteristics of specific crops (§ 2.2). The Commission also provided an "indicative catalogue" of measures, including on-farm practices, that might form part of a Member State's coexistence strategy (§ 3).

Though the Commission rejected prohibition of GM crops, it encouraged voluntary cooperation among farmers. Farmers could agree to establish zones of a single production type to reduce costs of crop segregation. Alternatively, producers could cluster fields with similar crop varieties, plant varieties with different flowering times, use different sowing dates, or coordinate crop rotations (§ 3.3).

### Reactions to the Recommendation

Not everyone agrees that coexistence is primarily an economic issue or that nonbinding guidelines are the most effective approach. The influential Economic and Social Committee asserted that coexistence affects agricultural land management, the food sector, land use and economic development, consumer protection, and environmental protection; it is closely connected with successful implementation of the entire scheme of GM regulation (EcoSoc 2005: 156–157). Therefore, coexistence "cannot be limited to the economic aspects of cultivation alone, but is an integral part of the risk management and prevention laid down by law" (157). Similarly, a legal analysis of the Commission Recommendation, commissioned by nongovernmental organizations, concluded (among other criticisms) that the focus on economic aspects is too restrictive, because Member States must also protect human health and the environment (Lasok and Haynes 2005).

The European Parliament issued a resolution that demanded both "uniform and binding rules to be established without delay at Community level" and Member State legislative measures to safeguard coexistence. If necessary, Member States should be able to prohibit cultivation of GMOs in some regions. Parliament also called for E.C.-wide civil liability and insurance requirements and would demand proof of insurance as part of the GMO authorization procedure. Indeed, Parliament asked that no new GM plants be authorized until binding rules on coexistence and a system of liability have been established (European Parliament 2003).

Changes in E.C. coexistence requirements remain possible. In June 2005, the Commission established a network group of Member State experts to exchange information about scientific studies and best practices and to facilitate the Commission's own study of coexistence (Commission Decision 2005/463). In March 2006, the Commission issued its report on the implementation of Member

State coexistence measures and noted that most States are still developing their regulatory frameworks (COM(2006)104). Support for E.C.-wide coexistence legislation seems to be growing, but by March 2007, no uniform E.C. rules had been proposed. Unless E.C. measures are imposed, Member States will govern coexistence.

### Member State Measures

Both Directive 2001/18 (art. 26) and the Commission Recommendation direct each Member State to enact measures to address coexistence for its own agricultural conditions. By the end of 2005, five States had enacted coexistence measures, and most others had published draft regulations. State requirements vary significantly and include on-farm practice requirements (training, duty to inform neighbors, record keeping), isolation distances for specific crops, liability provisions, compensation funds, and other measures (COM(2006)104).

In June 2004, Denmark enacted its Act No. 436 on the Growing etc. of Genetically Modified Crops (the Act on Co-existence), which authorizes the Minister for Food, Agriculture, and Fisheries to make rules to manage the coexistence of GM and other crops (Denmark Act No. 436, §§ 3–8). Rules may require a license for growing, handling, and transport of GM crops, issue restrictive authorizations for growing GM crops, and limit sales of GM materials to authorized growers (§§ 3–6). The Minister may impose other obligations, including notification of GM crops to nearby owners and producers, prescribed separation distances, and reports on field locations (§ 6). The Minister also has authority to pay compensation, within limits, to farmers who suffer losses from unintended GM material in crops, with funds collected through an annual fee per hectare of GM crops (§§ 9–12). A Ministerial Order enacted in 2005 establishes rules, including training for producers, reporting of GM fields, payment of 100 Danish kroner (about $18.00 U.S.) per GM hectare into the compensation fund, and other measures. Conditions of cultivation specify cultivation distances, crop-year intervals between GM and organic planting, seed handling, and notice requirements for maize, beets, and potatoes (Denmark Ministerial Order 2005).

In November 2005, the European Commission authorized Denmark's compensation scheme, financed with funds from GM producers. Farmers who suffer economic loss from admixture of conventional or organic crops, with GM material exceeding 0.9%, may receive compensation, limited to the difference in price between crops labeled "GM" and conventional or organic varieties. The Danish government will recover costs from the responsible GM farmer. The Commission approved the fund for five years, with the expectation that private insurance will eventually be available (European Commission 2005b).

Germany's Genetic Modification Act, passed in 2004, prescribes stringent coexistence measures. Farmers who plan to cultivate GM crops (and operators who release a GMO) must notify federal authorities, who will include the information in a site register accessible to the public. Those who cultivate GM products must take precautionary action to avoid admixture and other adverse effects. Rules will specify good farming practice, which may include minimum distances, variety selection, and record keeping. Significantly, farmers who cultivate GM crops are liable for "material adverse effects" that occur if, because of GM "contamination," products cannot be marketed, can be marketed only if they are labeled GM, or cannot be marketed with a label (e.g., organic) that would have been possible but for contamination. Moreover, if several producers have caused the adverse effects and the damage cannot be apportioned, each producer can be held responsible (Germany, Gentechnikgesetz 2005: arts. 1, §§ 16a, 16b, 36a; see Smyth and Kershen 2006). The strict liability provision, with joint and several liability, is likely to discourage German producers from planting GM crops. German legislators are likely to consider further amendments to the Genetic Modification Act.

Italy's coexistence law (Legge 2005) indicated that conventional, organic, and GM farming should coexist, but that introduction of transgenic crops must occur without the "slightest prejudice" to existing farms and without requiring changes in normal farming techniques (art. 2). The law established an expert advisory committee for coexistence (art. 7), charged with formulating guidelines for national framework regulations for coexistence. Thereafter, regions would adopt coexistence plans with technical rules and possibly compensation funds to restore damaged agricultural land; plans should encourage farmers to enter voluntary agreements for management measures to ensure coexistence (art. 4). Until regional rules were adopted, transgenic crops would not be permitted (arts. 6, 8). Once regional plans were adopted, farmers would be required to follow a coexistence plan, based on the regional plan; failure would result in liability for damages (art. 5).

Because many Italian regions oppose GM crops (e.g., Tuscany and Umbria have prohibited use of GM plants), regions were not eager to end the temporary moratorium by adopting plans. In fact, the Regional Council of Marche (a GMO-free region) challenged the law. In March 2006, the Italian Constitutional Court held that the coexistence law is unconstitutional. The Court upheld only the first two articles of the law, which define transgenic, organic, and conventional farming (art. 1) and safeguard the principle of coexistence (art. 2). The rest of the law interfered with the competence of Italy's 20 regions and was therefore unconstitutional (Italy, Constitutional Court 2006). Italian regions must now regulate coexistence; regional bans remain against E.C. law (see *Land Oberösterreich*).

In the Netherlands, the government believes that management of coexistence, an economic question, should occur through agreements between those who cultivate GM and non-GM crops. Therefore, a Committee on Coexistence in the Primary Sector, with representatives from major farmers' and plant breeders' organizations, negotiated a voluntary agreement. The agreement specifies separation distances for GM maize, potatoes, and sugar beets; distances are greater when neighbors have a registered GM-free or organic crop. Under the agreement, farmers who plan to grow GM crops must register before February 1 of each crop year. They must notify neighboring growers and those within separation distances; neighbors who market GM-free crops must notify the GM grower, who must then observe increased separation distances. Farmers who comply with separation distances and other requirements avoid liability, and a fund—with contributions from all parties in the food chain for each crop—will compensate those who suffer economic damage (Netherlands, Committee on Co-existence 2004). The Dutch Commodity Board for Arable Farming implemented the agreement in a 2005 regulation, which obligates producers to comply with notification provisions, crop separation distances, and the duty to separate GM and non-GM crops (Netherlands, Commodity Board 2005).

Member State coexistence measures continue to develop, as other States and some regions submit draft measures to the Commission. Though it is too early to generalize, a few observations can be made. As they develop measures, States have attempted to consult with stakeholders and to base their measures on scientific research (COM(2006)104: 5–6). States agree that educating producers about farming practices is essential, and some require training courses. In some States, producers of GM crops must register to inform the public (or only specific interested parties) about their production. Separation distances help to avoid pollen drift, and these vary by State and by crop. Member States with coexistence measures take various approaches to liability, from strict liability (Germany) to fault-based liability with a compensation fund if no producer is at fault (Netherlands), to a more general compensation fund, with government to recover from the GM farmer (Denmark). Experience in Member States will help to guide the European Community, as policy makers decide whether E.C. regulation of coexistence should be enacted.

## Liability

### Preliminary Considerations

The issue of liability is closely related to coexistence, and the assignment of liability for damages that might arise from development or use of GMOs raises difficult issues, both for the European Community and for Member States. European Community measures that govern GM crops and their products do not address liability comprehensively, and more general measures apply only in limited circumstances. As the discussion above indicated, some States have addressed liability for economic damage in their coexistence legislation.

Directive 2001/18 governing deliberate release does not address liability, and its preamble indicates that the Directive is without prejudice to national legislation on environmental liability. Measures to assign liability for environmental damage were to be enacted in a more general liability scheme for GMOs and other dangerous activities (preamble (16)).

Moreover, the Commission Recommendation on coexistence insisted that Member States should address liability for GM crops:

> The type of instruments adopted may have an impact on the application of national liability rules in the event of economic damage resulting from admixture. Member States are advised to examine their civil liability laws to find out whether the existing national laws offer sufficient and equal possibilities in this regard. Farmers, seed suppliers and other operators should be fully informed about the liability criteria that apply in their country in the case of damage caused by admixture. (Commission Recommendation 2003/556: Annex § 2.1.9)

The Recommendation suggested that an insurance scheme might help to compensate damage from admixture of GM and other crops (Annex § 2.1.9). Some Member State coexistence measures encourage or require insurance, although insurance for "economic damage resulting from adventitious presence is not available in the EU" (COM(2006)104: 7).

The E.C. Products Liability Directive (Council Directive 85/374, as amended), which applies broadly, prescribes that "[t]he producer shall be liable for damage caused by a defect in his product" (art. 1). A product is defective when "it does not provide the safety which a person is entitled to expect, taking all circumstances into account" (art. 6). To help restore consumer confidence in safety of agricultural products, the Directive was amended in 1999 to include primary agricultural products. But the requirement that the product be defective (questionable for authorized GMOs) suggests that measures enacted under the Products Liability Directive will rarely redress damage from GMOs (art. 9) (Grossman 2003: 97; see also Bergkamp 2003: 252–258).

Some argue that no special liability regime for GMOs is justified because regulation has reduced risk to "acceptable levels," and both "scientific knowledge and practical experience" indicate that no significant, unreasonable risk remains (Bergkamp 2000: 110). Even when policy makers agree that a liability scheme might be desired, the design of such a scheme fosters disagreement. For example, negotiations prior to enactment of the Cartagena Protocol on Biosafety considered numerous alternative liability regimes for biotechnology, but no agreement could be reached. Instead, the Protocol (art. 27) calls for further study (see Smyth and Kershen 2006: 50–77). Nor do other liability regimes (e.g., the Council of Europe's Lugano Convention, not yet in force) impose liability on agricultural producers.

Some national liability regimes provide remedies through laws imposing strict liability for defective goods or substances or through fault-based principles of negligence or nuisance (Bergkamp 2000: 66–70, 104–107; Grossman 2003: 97). Specifically for GMOs, coexistence measures have begun to address producer liability, albeit inconsistently. For example, German law (discussed above) imposes strict liability on producers who cause a material adverse effect on other property owners, whereas other coexistence laws authorize compensation systems and impose only fault-based liability.

### Environmental Liability Directive: Directive 2004/35

The Commission's *White Paper on Environmental Liability* reflects public concerns that GMOs may cause environmental harm (COM(2000)66). The *White Paper* recommended that liability for damage from GMOs be treated in a general framework measure addressing liability in a number of sectors. That measure is Directive 2004/35 on environmental liability with regard to the prevention and remedying of environmental damage. Invoking the polluter pays principle and the principle of sustainable development (art. 1), the Directive governs only environmental damage, rather than traditional damage to persons and property (preamble (14): art. 3).

The Directive defines environmental damage to include harm to protected species and natural habitats—generally those protected by the Wild Birds and Habitats Directives (Council Directives 79/409 and 92/43) or by national nature conservation legislation. It also includes water damage and land damage (Directive 2004/35: art. 2(1)). The latter is contamination that creates a significant risk to human health from "direct or indirect introduction, in, on or under land, of substances, preparations, organisms, or micro-organisms" (art. 2(1)(c)). The Directive does not give private parties a right to compensation for environmental damage, though Member States may do so (art. 3(3)).

By 30 April 2007, Member States must enact national measures to comply with the Directive, which does not apply to damage caused before that date (arts. 17, 19(1)). Member State measures must require operators (those who carry out, or hold authorization for, the listed activity) to take preventive action to avoid environmental damage, to apply measures to remediate the damage, and to bear costs for preventive and remedial actions (arts. 2(6), 6–8).

The Directive applies to environmental damage from specific activities, including contained use of GMMs and deliberate release and marketing of GMOs (art. 3(1), Annex III). Thus, operators have the duty to prevent and to remediate damage from GMOs. But Member States may allow operators not to bear the cost of remedial actions under some conditions—if the operator was not at fault or negligent and the damage was caused by an authorized emission or event in compliance with applicable law or by an emission or activity that the operator can show was not considered likely to cause environmental damage "according to the state of scientific and technical knowledge" when the emission or activity took place (art. 8(4)). This exemption would seem to apply to authorized GMOs used in accordance with their authorizations. Member States can maintain or adopt more stringent measures to prevent and remedy environmental damage and can identify additional activities and responsible parties (art. 16).

Because of the intentionally limited scope of the Directive, Member State legislation will continue to apply to redress traditional damage to persons and property. Thus, generally applicable tort measures or specific coexistence legislation will play a role in allocating liability for possible damage from GMOs.

## CONCLUDING OBSERVATIONS

The major components of the updated E.C. regulatory system for GM crops, food, and feed are now in effect, and regulators have authorized several new products under these measures. Nonetheless, Member States remain divided about GMOs, and fears about their impact on health and the environment continue, even after successful risk assessments. These Member State divisions and fears may help to explain repeated delays in approval of new GM products and bans imposed by Member States, regions, and even municipalities.

Because imported products must comply with E.C. requirements, E.C. regulatory measures exert significant influence beyond the borders of the European Union and its 27 Member States. E.C. reluctance to approve new products limits trade by preventing import of GMOs, still unauthorized in Europe, that have been grown and consumed without incident in the United States and elsewhere. The 2003 complaint against the European Communities (the name used in WTO matters), filed by the United States, Canada, and Argentina under the World Trade Organization (WTO) dispute settlement system, formalizes allegations that the European Communities and its Member States violated WTO requirements (see Grossman 2005: 80–83). The September 2006 panel report, accepted by the Dispute Settlement Body in November 2006, found that the European Communities had applied a de facto moratorium on approval of GM products and therefore violated its WTO obligations. In addition, both the undue delay associated with approval of specific GM products and Member State safeguard measures (i.e., bans) violated WTO obligations (WTO 2006). Challenges to E.C. labeling requirements under the WTO remain possible.

Even when GM crops, food, and feed have been approved in the European Community, regulations impose significant burdens on U.S. food and feed companies and producers. The European Community's "double requirement for mandatory biotechnology-specific labeling and biotechnology-specific traceability" is particularly burdensome (USDA 2005: 2). To ensure that conventional products do not trigger thresholds for labeling, agribusiness firms must ensure segregation of GM and traditional crops in the complex grain-handling system. Though some firms already have systems for traceability, others must develop and implement them. These requirements reduce "the fungibility of the product and the flexibility of the production systems" (2). Moreover, they are costly. As a USDA committee observed, "[t]he greater the requirements for documentation and segregation from the commodity stream, the greater the costs associated with originating and manufacturing" particular food or feed ingredients (9). The requirement that food and feed products be labeled, even when GM protein and DNA cannot be detected (e.g., in highly refined products), imposes additional burdens (12–13).

European Community traceability and labeling measures affect U.S. farmers, as well, imposing additional costs and risks and forcing changes in production practices. Some farmers will plant only traditional crops for export but risk admixture with GM varieties and the resulting economic consequences. Those who grow GM varieties must segregate GM from other crops. Production practices that minimize pollen drift and other causes of admixture are important for both traditional and GM varieties. The low E.C. threshold for labeling products with adventitious or technically unavoidable presence of GMOs, with an even lower threshold possible for seed, makes avoidance of admixture critical.

Both traceability and labeling are consistent with a regulatory system based on a strict view of the precautionary principle and influenced, perhaps, by consumer fears—whether justified or not—about the safety of food products. To be sure, labeling of GMOs facilitates consumer choice by informing consumers of GM content above the regulatory threshold. But labeling may also stigmatize GM products and discourage consumption of foods with no known health risks (Mansour and Key 2004: 64–68; Appleton 2000).

Different U.S. and E.U. attitudes toward GM food and feed on the part of citizens, the agricultural community, and regulatory agencies help to account for differences in regulatory requirements for GM crops and their products. The European approach, with rules based on process and heavily influenced by the precautionary principle, is unique in its "double requirement" of labeling and traceability (USDA 2005: 13). Though GM varieties are a significant percentage of U.S. agricultural production, U.S. law requires neither traceability nor labeling. Thus, E.C. rules impose additional obligations—labeling or proving that products are exempt from labeling requirements—for U.S. producers. It is to be hoped that incompatibilities in the U.S. and E.C. regulatory systems can be resolved without further damage to international trade and to U.S.–E.U. relations.

## Notes

This chapter is based on work supported by the Cooperative State Research, Education and Extension Service, U.S. Department of Agriculture, under Project No. ILLU-470-309. It is based, with permission, on Grossman's article published in the *Journal of Food Law and Policy* (Grossman 2005), which won the 2006 Professional Scholarship Award from the American Agricultural Law Association.

1. The European Community, established by the Treaty of Rome in 1957, is the current name for the European Economic Community, one of three Communities designed to facilitate economic integration. The three Communities are the basis of the European Union, formed by the Treaty on European Union, effective 1 November 1993. The European Community, which enacts regulatory measures under authority of the Treaty of Rome, as amended, is governed by EU institutions, including the European Council, Parliament, and Commission. In WTO matters, the European Union is known as the European Communities.

2. The approval for GT 73 had been published in June 2005 (2005 O.J. (L 164) 57), but was immediately withdrawn (2005 O.J. (L 165) 34).

3. In June 2004, E.U. leaders agreed on the text of a new Treaty establishing a Constitution for Europe. The Constitution would create one Union to replace the European Communities and European Union. The heads of state and government of the Member States signed the Constitution in October 2004, and all then-25 States should have ratified the Constitution before it could enter into force (Treaty Establishing a Constitution for Europe 2004). In spring 2005, however, both the Netherlands and France rejected the Constitution in national referenda, raising doubts about its success.

4. Because the Commission decision was addressed to the Republic of Austria, and not to Upper Austria, the court also considered Upper Austria's standing to sue. The court concluded that Upper Austria had standing; it was individually affected by the Commission decision that concerned its draft law (*Land Oberösterreich:* ¶¶ 25–30). The decision also considered and rejected other claims, including breach of the right to be heard and the obligation to state reasons and violation of the Precautionary Principle (¶ 32–58, 70–73).

5. Compare a narrower definition preferred by U.S. producers and firms: "The efficient and rapid tracking of physical product and traits from and to critical points of origin or destination in the food chain necessary to achieve specific food safety and, or, assurance goals" (Farm Foundation 2004: 9).

6. The European Court of Justice considered the application of the labeling threshold from a Regulation (1139/98, which applied to certain GM soy and maize), now replaced by Regulations 1829/2003 and 1830/2003. In a case referred from Italy, the Court held that the labeling threshold (under that regulation, 1%) applied to foodstuffs intended for

the use of infants and young children. Risk assessment of GMOs was intended to ensure that authorized products are safe for the consumer, so no justification exists for applying a stricter standard to label foods for infants (*Ministero della Salute v. Codacons* 2005).
7. A practical example of the issue was raised in a case decided in the United Kingdom, *Regina v. Secretary of State* (1998). Watson, a major organic grower in Britain, feared cross-pollination between his crop and GM trial plantings on a neighboring farm. The court refused to order destruction or detasseling because risk of cross-pollination was slight and the GM producer had consent for trial planting.

## References

### European Community Legislative Measures

Treaties

E.C. Treaty (Treaty Establishing the European Community), as amended. Consolidated version at 2002 O.J. (C 325) 33.

Treaty Establishing a Constitution for Europe. 16 Dec. 2004. 2004 O.J. (C 310) 1. (Not in force)

Secondary Legislation

Commission Decision 2002/623 establishing guidance notes supplementing Annex II to Directive 2001/18, 2002 O.J. (L 200) 22.

Commission Decision 2003/653 relating to national provisions on banning the use of genetically modified organisms in the region of Upper Austria, 2003 O.J. (L 230) 34.

Commission Decision 2003/701 establishing pursuant to Directive 2001/18 ... a format for presenting the results of the deliberate release into the environment of genetically modified higher plants for purposes other than placing on the market, 2003 O.J. (L 254) 21.

Commission Decision 2004/204 laying down detailed arrangements for the operation of the registers for recording information on genetic modifications in GMOs, provided for in Directive 2001/18, 2004 O.J. (L 65) 20.

Commission Decision 2004/643 concerning the placing on the market, in accordance with Directive 2001/18 ... of a maize product (*Zea mays L.* line NK 603) genetically modified for glyphosate tolerance, 2004 O.J. (L 295) 35.

Commission Decision 2004/657 authorising the placing on the market of sweet corn from genetically modified maize line Bt11 as a novel food or novel food ingredient under Regulation No. 258/97, 2004 O.J. (L 300) 48.

Commission Decision 2005/448 authorising the placing on the market of foods and food ingredients derived from genetically modified maize line NK 603 as novel foods or novel food ingredients under Regulation No. 258/97, 2005 O.J. (L 158) 20.

Commission Decision 2005/463 establishing a network group for the exchange and coordination of information concerning coexistence of genetically modified, conventional and organic crops, 2005 O.J. (L 164) 50.

Commission Decision 2005/608 concerning the placing on the market, in accordance with Directive 2001/18 ... of a maize product (*Zea mays* L., line MON 863) genetically modified for resistance to corn rootworm, 2005 O.J. (L 207) 17.

Commission Decision 2005/635 concerning the placing on the market, in accordance with Directive 2001/18 ... of an oilseed rape product (*Brassica napus* L., GT73 line) genetically modified for tolerance to the herbicide glyphosate, 2005 O.J. (L 228) 11.

Commission Decision 2005/772 concerning the placing on the market, in accordance with Directive 2001/18 ... of a maize product (*Zea mays* L., line 1507) genetically modified for resistance to certain lepidopteran pests and for tolerance to the herbicide glufosinate-ammonium, 2005 O.J. (L 291) 42.

Commission Recommendation 2003/556 on guidelines for the development of national strategies and best practices to ensure the coexistence of genetically modified crops with conventional and organic farming, 2003 O.J. (L 189) 36.

Commission Recommendation 2004/787 on technical guidance for sampling and detection of genetically modified organisms and material produced from genetically modified organisms as or in products in the context of Regulation (E.C.) No. 1830/2003, 2004 O.J. (L 348) 18.

Commission Recommendation 2005/637 concerning the measures to be taken by the consent holder to prevent any damage to health and the environment in the event of accidental spillage of an oilseed rape, 2005 O.J. (L 228) 19.

Commission Regulation 65/2004 establishing a system for the development and assignment of unique identifiers for genetically modified organisms, 2004 O.J. (L 10) 5.

Commission Regulation 641/2004 on detailed rules for the implementation of Regulation 1829/2003 as regards the application for the authorisation of new genetically modified food and feed, the notification of existing products and adventitious or technically unavoidable presence of genetically modified material which has benefited from a favourable risk evaluation, 2004 O.J. (L 102) 14.

Council Decision 1999/468 laying down the procedures for the exercise of implementing powers conferred on the Commission, 1999 O.J. (L 184) 23. Amended from July 2006 by Council Decision 2006/512, 2006 O.J. (L 200) 11.

Council Decision 2002/811 establishing guidance notes supplementing Annex VII to Directive 2001/18, 2002 O.J. (L 280) 27.

Council Decision 2002/812 establishing pursuant to Directive 2001/18 ... the summary information format relating to the placing on the market of genetically modified organisms as or in products, 2002 O.J. (L 280) 37

Council Decision 2002/813 establishing, pursuant to Directive 2001/18 ... the summary notification information format for notifications concerning the deliberate release into the environment of genetically modified organisms for purposes other than for placing on the market, 2002 O.J. (L 280) 62.

Council Directive 79/409 on the conservation of wild birds, 1979 O.J. (L 103) 1, as amended.

Council Directive 85/374 on the approximation of the laws, regulations and administrative provisions of the Member States concerning liability for defective products, 1985 O.J. (L 210) 29, as amended by Parliament and Council Directive 1999/34, 1999 O.J. (L 141) 20.

Council Directive 90/219 on the contained use of genetically modified micro-organisms, 1990 O.J. (L 117) 1, as amended. Consolidated version at CONSLEG 1990L0219, 20/11/2003.

Council Directive 90/220 on the deliberate release into the environment of genetically modified organisms, 1990 O.J. (L 117) 15) (repealed effective 17 Oct. 2002, when Council Directive 2001/18 took effect).

Council Directive 92/43 on the conservation of natural habitats and of wild fauna and flora, 1992 O.J. (L 206) 7, as amended.

Council Directive 2000/13 on the approximation of the laws of the Member States relating to the labelling, presentation and advertising of foodstuffs, 2000 O.J. (L 109) 29, as amended.

Council Directive 2002/53 on the common catalogue of varieties of agricultural plant species, 2002 O.J. (L 193) 1, as amended. Consolidated version at CONSLEG 2002L0053, 18/04/2004.

Directive 2001/18 of the European Parliament and of the Council on the deliberate release into the environment of genetically modified organisms and repealing Council Directive 90/220, 2001 O.J. (L 106) 1, as amended. Consolidated version at CONSLEG 2001L0018, 07/11/2003.

Directive 2004/35 of the European Parliament and of the Council on environmental liability with regard to the prevention and remedying of environmental damage, 2004 O.J. (L 143) 56.

Draft Commission Decision establishing minimum thresholds for adventitious or technically unavoidable traces of genetically modified seeds in other products, www.genfood.at/download/com_draft_seeds_04_2004.pdf.

Regulation 258/97 of the European Parliament and of the Council concerning novel foods and novel food ingredients, 1997 O.J. (L 43) 1 (repealed as to GM foods and ingredients).

Regulation 178/2002 [General Food Law] of the European Parliament and of the Council of 28 January 2002 laying down the general principles and requirements of food law, establishing the European Food Safety Authority and laying down procedures in matters of food safety, 2002 O.J. (L 31) 1.

Regulation 1829/2003 of the European Parliament and of the Council on genetically modified food and feed, 2003 O.J. (L 268) 1.

Regulation 1830/2003 of the European Parliament and of the Council concerning the traceability and labelling of genetically modified organisms and the traceability of food and feed products produced from genetically modified organisms and amending Directive 2001/18, 2003 O.J. (L 268) 24.

### Commission Documents

COM(2000)1 final. Communication from the Commission on the Precautionary Principle.

COM(2000)66 final. White Paper on Environmental Liability.

COM(2004)250 final. Life Sciences and Biotechnology—A Strategy for Europe: Second Progress Report and Future Orientations.

COM(2004)575 final. Report from the Commission to the Council and the European Parliament on the experience of member states with GMOs placed on the market under Directive 2001/18/EC and incorporating a specific report on the operation of parts B and C of the directive.

COM(2006)104 final. Communication from the Commission to the Council and the European Parliament. Report on the implementation of national measures on the coexistence of genetically modified crops with conventional and organic farming.

### Other Legislative Measures

Cartagena Protocol on Biosafety to the Convention on Biological Diversity (Montreal 2000), 39 I.L.M. 1027.

Convention on Biological Diversity, U.N. Conference on Environment and Development, 5 June 1992, 31 I.L.M. 818.

Denmark, Act No. 436 of 9 June 2004 on the Growing etc. of Genetically Modified Crops (E.C. TRIS notification no. 2004/393/DK).

Denmark, Ministerial Order of 31 March 2005 on the Cultivation etc. of Genetically Modified Crops (E.C. TRIS notification no. 2004/546/DK).

Germany, Gentechnikgesetz. 3 Feb. 2005. Gesetz zur Neuordnung des Gentechnikrechts vom 21 Dezember 2004. Bundesgesetzblatt 2005 I nr. 8:186–196.

Italy, Legge. 28 gennaio 2005, n. 5. Gazzetta Ufficiale, serie generale, n. 22 (most provisions held unconstitutional in judgment no. 116 of 2006). An English translation is available in USDA, Foreign Agricultural Service, GAIN Report No. IT5003.

Netherlands, Committee on Co-existence. 2004. Co-existence in the Primary Sector.

Netherlands, Commodity Board for Arable Farming. 2005. HPA Regulation on the Coexistence of Crops (JBA No. 322, E.C. TRIS notification no. 2006/97/NL).

Rio Declaration on Environment and Development, U.N. Conference on Environment and Development. U.N. Doc. 4/conf.151/26 (vol. 1), U.N. Sales No. E.73.II.A.14 (1992).

## Court and Panel Decisions

Association Greenpeace France v. Ministère de l'Agriculture et de la Pêche, Case C-6/99 (2000), ECR I-1651.

Commission v. French Republic, Case C-296/01 (2003), ECR I-13909.

Italy, Constitutional Court of the Italian Republic. 8 Mar. 2006. Judgment No. 116.

Land Oberösterreich v. Commission, Cases T-366/03 & T-235/04, Court of First Instance, Judgment of 5 October 2005, curia.europa.eu. [Case C-492/03 (renumbered T-235-04), 2004 O.J. (C 21) 20; Case T-366/03, 2004 O.J. (C 35) 11.]. Appeals are noted at 2006 O.J. (C 48) 15 & 2006 O.J. (C 60) 20; Cases C-439/05P & C-454/05P, consolidated by order of 29 June 2006.

Ministero della Salute v. Codacons, Case C-132/03, Judgment of 26 May 2005, curia.europa.eu.

Regina v. Secretary of State for the Environment *ex parte* Watson (1998), EWCA Civ. 1250 (21 July 1998), bailii.org/ew/cases/EWCA/Civ/1998/1250.html.

WTO. 29 Sept. 2006. Panel Report, European Communities—Measures Affecting the Approval and Marketing of Biotech Products. WT/DS291–293/R. Accepted by the Dispute Settlement Body, Doc. WT/DS 291/33, 29 Nov. 2006.

## Secondary Literature

Appleton, Arthur E. 2000. The Labeling of GMO Products Pursuant to International Trade Rules. *New York University Environmental Law Journal* 8:566–577.

Bergkamp, Lucas. 2000. Allocating Unknown Risk: Liability for Environmental Damages Caused by Deliberately Released Genetically Modified Organisms. *Tijdschrift voor Milieuaansprakelijkheid* 2000:61–70 (part 1), 104–114 (part 2).

Bergkamp, Lucas. 2003. *European Community Law for the New Economy*. Antwerp, Belgium: Intersentia.

Biosafety Clearing-House. Central Portal. bch.biodiv.org/.

Brosset, Estelle. 2004. The Prior Authorisation Procedure Adopted for the Deliberate Release into the Environment of Genetically Modified Organisms: The Complexities of Balancing Community and National Competences. *European Law Journal* 10:555–579.

Buenderman, Mark. 5 Oct. 2005. Regions Not Allowed to Ban GMOs, Court Rules. *EU Observer*.

de Sadeleer, Nicolas. 2002. *Environmental Principles: From Political Slogans to Legal Rules*. Oxford: Oxford University Press.

Directorate-General for Agriculture and Rural Development. 2005. *Agriculture in the European Union—Statistical and Economic Information 2004.*

Directorate-General for Agriculture and Rural Development. 2006. *Agriculture in the European Union—Statistical and Economic Information 2005.*

EcoSoc. 2005. Opinion of the European Economic and Social Committee on the Co-existence between Genetically Modified Crops, and Conventional and Organic Crops. 2005 O. J. (C 157) 155.

EFSA (European Food Safety Authority). 2003. Opinion of the Scientific Panel on Genetically Modified Organisms ... for the placing on the market of herbicide-tolerant genetically modified maize NK 603, for import and processing. *E.F.S.A. Journal* 10:1–13.

EFSA (European Food Safety Authority). 2004a. Opinion of the Scientific Panel on Genetically Modified Organisms on a request from the Commission related to the Austrian invoke of Article 23 of Directive 2001/18/EC. *E.F.S.A. Journal* 78:1–13.

EFSA (European Food Safety Authority). 2004b. Opinion of the Scientific Panel on Genetically Modified Organisms on a request from the Commission related to the Greek invoke of Article 23 of Directive 2001/18/EC. *E.F.S.A. Journal* 79:1–8.

EFSA (European Food Safety Authority). 2004c. Guidance Document of the Scientific Panel on Genetically Modified Organisms for the Risk Assessment of Genetically Modified Plants and Derived Food and Feed. *E.F.S.A. Journal* 99:1–94.

EuropaBio. 2007. Green Biotechnology Manifesto. www.greenbiotech-manifesto.org/screen-manifesto-900.pdf

European Commission. 5 Mar. 2003. Press Release, GMOs: Commission Addresses GM Crop Co-existence (IP/03/314).

European Commission. 2004a. Guidance on the Implementation of Articles 11, 12, 16, 17, 18, 19 and 20 of Regulation (E.C.) No. 178/2002 on General Food Law. ec.europa.eu/food/food/foodlaw/guidance/index_en.htm.

European Commission. 2004b. Plants for the Future—2025: A European Vision for Plant Genomics and Biotechnology. Brussels: E.U. Publications Office.

European Commission. 5 Nov. 2004c. Press Release, Dioxin Contamination: EU Traceability and Alert Notification Systems Work Well (IP/04/1343).

European Commission. 8 Sept. 2004d. Press Release, Inscription of MON 810 GM Maize Varieties in the Common EU Catalogue of Varieties (IP/04/1083).

European Commission. 2005a. Invocation of Article 16 under Directive 90/220/EC and Article 23 under Directive 2001/18/EC (Safeguard clause) as of 15 Mar. 2005. ec.europa.eu/comm/environment/biotechnology/safeguard_clauses.htm.

European Commission. 2005b. State Aid No. 568/2004, Denmark. ec.europa.eu/community_law/eulaw/index_en.htm#aides (see Press Release IP/05/1458, 23 Nov. 2005).

European Commission. 2006. List of the Genetically Modified Material Which Has Benefited from a Favourable Risk Evaluation within the Meaning of Article 47 of Regulation (E.C.) No 1829/2003 (last updated 7 Feb. 2006). ec.europa.eu/food/food/biotechnology/gmfood/events_en.pdf.

European Commission. 27 Mar. 2007. Press Release, GMOS: Three oilseed rapes authorised for import and processing in animal feed (IP/07/416).

European Council. 26 Apr. 2004. Press Release, 2578th Council Meeting, Agriculture and Fisheries (IP/8350/04).

European Parliament. 2003. Resolution on coexistence between genetically modified crops and conventional and organic crops (2003/2098(INI)). 2004 O.J. (C 90 E) 680.

Farm Foundation. 2004. *Food Traceability & Assurance in the Global Food System.* Oakbrook, IL: Farm Foundation.

Fischler, Franz. 25 Feb. 2003. Communication from Mr. Fischler to the Commission, Co-existence of Genetically Modified, Conventional and Organic Crops.

Forsman, Zeynep Kivilcim. 2004. Community Regulation of Genetically Modified Organisms: A Difficult Relationship Between Law and Science. *European Law Journal* 10:580–594.

Francescon, Silvia. 2001. The New Directive 2001/18/EC on the Deliberate Release of Genetically Modified Organisms into the Environment: Changes and Perspectives. *Review of European Community and International Environmental Law* 10:309–320.

Grossman, Margaret Rosso. 1995. Agro-environmental Measures in the Common Agricultural Policy. *University of Memphis Law Review* 25:927–1049.

Grossman, Margaret Rosso. 2003. Genetically Modified Crops in the United States: Federal Regulation and State Tort Liability. *Environmental Law Review* 5:86–108.

Grossman, Margaret Rosso. 2005. Traceability and Labeling of Genetically Modified Crops, Food, and Feed in the European Union. *Journal of Food Law and Policy* 1:43–85.

Grossman, Margaret Rosso, & A. Bryan Endres. 2000. Regulation of Genetically Modified Organisms in the European Union. *American Behavioral Scientist* 44:378–434 (2000).

James, Clive. 2006. Global Status of Commercial Biotech/GM Crops: 2006. ISAA Brief 35-2006: Executive summary. www.isaaa.org/resources/publications/briefs/35/executivesummary/default.html.

Lasok, K. P. E., & Rebecca Haynes. 21 Jan. 2005. In the Matter of Co-existence, Traceability and Labelling of GMOs: Advice. www.foe.co.uk/resource/briefings/legal_opinion_in_the_matte.pdf.

MacMaoláin, Caoimhín. 2003. The New Genetically Modified Food Labelling Requirements: Finally a Lasting Solution. *European Law Review* 28:865–879.

Mansour, Mark, & Sarah Key. 2004. From Farm to Fork: The Impact on Global Commerce of the New European Union Biotechnology Regulatory Scheme. *International Lawyer* 38:55–69.

OECD. The Biotech Database. webdomino1.oecd.org/ehs/bioprod.nsf.

Smyth, Stuart J., & Drew L. Kershen. 2006. Agricultural Biotechnology: Legal Liability Regimes from Comparative and International Perspectives. *Global Jurist Advances* 6(2): art. 3.

USDA Advisory Committee on Biotechnology and 21st Century Agriculture (AC21). 2005. Global Traceability and Labeling Requirements for Agricultural Biotechnology-Derived Products: Impacts and Implications for the United States. w3.usda.gov/agencies/biotech/ac21/reports/tlpaperv37final.pdf.

# 5

# Genetically Engineered Animals and the Ethics of Food Labeling

*Robert Streiffer & Alan Rubel*

Much of the current debate about labeling policy for genetically engineered (GE) food focuses exclusively on foods derived from GE crops and ignores the possibility that food would be, or would include as an ingredient, meat from GE animals ("GE animal products"). For example, Neil Belson's (2000) overview of U.S. regulations concerning agricultural biotechnology focuses entirely on GE plants. And although the website of the U.S. Animal and Plant Health Inspection Service (Animal and Plant Health Inspection Service n.d.) contains a summary of the U.S. agencies responsible for overseeing agricultural biotechnology, the summary does not address meat, poultry, or egg products.

The tendency to ignore GE animal products is not surprising since no GE animal products are currently available. Cheese made using recombinantly produced rennet and foods that contain ingredients derived from Bt corn or Roundup Ready soybeans constitute the vast majority of available GE foods, but these do not involve GE animals. Even in the case of recombinantly produced bovine growth hormone, a product frequently discussed under the heading of "animals and genetic engineering," genetic engineering is applied only to microorganisms, not animals. Participants in the debate may also have shown a tendency to ignore GE animal products because they have assumed that GE microorganisms and GE plants raise the same issues as GE animals and thus that conclusions drawn about the first two automatically extend to the third.

This tendency is unjustified for three reasons: (1) animals raise ethical issues beyond those raised by plants and microorganisms; (2) the U.S. regulatory system treats animal food products differently from other food products; and (3) two GE animals—the Enviropig and the AquAdvantage Bred salmon—are currently under review for commercialization in the United States. We suggest that it is time to extend the GE food labeling debate to encompass GE animal products.

The bulk of research on GE animal products has been conducted on cattle, sheep, goats, pigs, chickens, and several different species of fish and has focused on the same traits of interest in traditional breeding: increased feed efficiency, faster growth rates, different fat-to-muscle ratios, and pest and disease resistance (Harper et al. 2003: 47–57, 7). This research is intended to produce GE animals that can be raised commercially and sold as food.

In addition to animals that have been genetically engineered for agricultural purposes, animals or animal parts that have been genetically engineered for other purposes can also enter the food supply. Some animals are engineered to facilitate basic research or to create nonfood products such as pharmaceuticals, industrial proteins, or organs for xenotransplantation. High rates of mosaicism (i.e., varied distribution of the transgene within the organism) and variable expression rates result in research and commercial production facilities having surplus transgenic animals, and these animals could be sold for food. There is already one documented case in which the offspring of investigational transgenic pigs entered the food supply, although according to the researcher, the offspring were not themselves transgenic (Food and Drug Administration 2003). Where the product of a transgene is limited to certain parts of the animal's body or is safe for human consumption, there will be GE animal parts that might be sold for food. Finally, in the case of animals engineered to produce pharmaceuticals or other drugs or chemicals in their milk during lactation, male offspring will be unnecessary and could be sold for food (National Research Council 2004: 7).

The U.S. food regulatory system has different pathways for fish, poultry, other livestock (including cattle, sheep, goats, and pigs), and eggs. The oversight of labeling for GE animal products has not yet solidified, but the following pathways are most likely (Pew Initiative on Food and Biotechnology 2004a: 100–140). First, some labeling could be overseen by the FDA's Center for Food Safety and Applied Nutrition under the general food regulations of the Food, Drug, and Cosmetic Act of 1938 (FDCA). This is the most obvious pathway for GE fish products, because this corresponds to the existing pathway for conventional fish products. Second, some labeling could be overseen by the USDA's Food Safety and Inspection Service (FSIS) acting under the Poultry Products Inspection Act of 1957 (PPIA), the Egg Products Inspection Act of 1970 (EPIA), and the Federal Meat Inspection Act of 1906 (FMIA). Third, some labeling could be overseen by the FDA's Center for Veterinary Medicine (CVM) acting under the new animal drug provisions of the FDCA. The FDA has declared that the products of the transgenes within GE animals constitute new animal drugs and must be approved by the CVM before marketing (Food and Drug Administration Center for Veterinary Medicine n.d.). However, for reasons we give below, having the CVM oversee *consumer* food product labeling is unlikely. Thus, we focus on the two other administrative bodies and the four corresponding pathways, the FDCA, the PPIA, the EPIA, and the FMIA.

## THE CONSUMER AUTONOMY ARGUMENT

It is important to distinguish between different decision makers and regulatory pathways because different decision makers will be acting under different norms and constraints (Streiffer and Rubel 2004: 224–225). Administrative agencies within a democracy, for example, do not have direct democratic obligations to represent the will of the people in the way that Congress does, and, unlike Congress, agencies are constrained by their authorizing statutes. Similarly, an argument that the FDA has an obligation to mandate labeling will not necessarily imply that a manufacturer has an obligation to voluntarily label in the absence of a legal requirement.[1] Given the range of actors in a position to provide or require information, it is thus important to give reasons for labeling that are appropriate to the decision maker being addressed.

In this chapter, we argue that the relevant norms for the FDA and the USDA require that they adopt mandatory labeling policies on GE animal products. This conclusion is based on the following version of the *consumer autonomy argument*:

1. If it is within the discretion of an agency to adopt a certain policy, and if the policy is the only one that adequately respects the values underlying the relevant parts of the agency's enabling legislation, then the agency ought to adopt that policy (absent sufficient countervailing considerations).
2. It is within the discretion of the FDA and the USDA to require labels on GE animal products.
3. The value underlying the relevant parts of their enabling legislation, namely, the relevant labeling statutes, is the protection of consumer autonomy.
4. Hence, if requiring labels on GE animal products is the only policy that adequately protects consumer autonomy, then the FDA and the USDA ought to require the labels (absent sufficient countervailing considerations).
5. Requiring labels on GE animal products is the only policy that adequately protects consumer autonomy.
6. There are no sufficiently strong countervailing considerations.
7. Hence, the FDA and the USDA ought to require labels on GE animal products.

## DEMOCRATIC OBLIGATIONS OF ADMINISTRATIVE AGENCIES

The first premise of the consumer autonomy argument is based on the idea that an administrative agency within a democracy has an obligation to respect not just the letter, but also the underlying values, of its authorizing statutes. In creating an administrative agency, Congress delegates a range of decision-making power to a body that is better suited to make judgments about complicated and changing circumstances. That is, administrative agencies have discretion by design. Because an agency should not exercise its discretion arbitrarily, there must be principles to guide its decision making, and one such principle derives from the origins of the agency's authority. An administrative agency's legitimate authority derives from authorizing statutes created by a democratically legitimate legislature. The agency therefore has a prima facie democratic obligation to exercise its discretion in ways that further or, at the very least, adequately respect the values underlying those statutes; failure to do so runs counter to the agency's democratic legitimacy. What an agency must do to comply with this norm is determined by answers to the following questions. First, which policies fall within the agency's discretion under its authorizing statutes? Second, what values underlie the relevant statutes? And third, to what extent do the various policy options respect those values? If only one policy adequately respects the underlying values, the agency has a prima facie democratic obligation to adopt that policy.

That prima facie obligation may, however, be overridden by reasons for adopting an alternative policy. To determine whether the alternative policy should prevail, one must ask two additional questions: First, how rich in democratic legitimacy are the authorizing statutes themselves? Second, how significant are the countervailing considerations, in terms of efficiency, justice, or legitimacy, when compared to the democratic legitimacy of the authorizing statutes?

## IS IT WITHIN THE FDA'S AND THE USDA'S DISCRETION TO REQUIRE LABELS?

Both the FDA and the USDA have the discretion to require labeling on GE animal products because (1) the labeling provisions of the FDCA, EPIA, FMIA, and PPIA contain a requirement that labels not be misleading and (2) labels on GE animal products that do not disclose that the product is a GE animal product will be misleading.

### Misleadingness and the Labeling Statutes

The FDCA, FMIA, and PPIA explicitly prohibit labels that are "misleading in any particular" (21 U.S.C. § 343(a); 21 U.S.C. § 601(n)(1); 21 U.S.C. § 453 (h)(1)). The EPIA similarly requires that labels include any information necessary "to assure that [products] will not have false or misleading labeling" (21 U.S.C. §1036(a)).

### Misleadingness and GE Food

In the case of GE crops, the existing lack of positive labels is misleading to many consumers. In the United States, approximately 70% of processed foods contain GE ingredients (Genetically Engineered Organisms-Public Issues Education Project 2004). Nonetheless, in 2003, 58% of U.S. consumers believed they had never eaten GE foods (Pew Initiative on Food and Biotechnology 2003b). In 2004, 41% were unsure whether GE foods were available in supermarkets, and 11% believed that they were not; 46% were unsure whether they had eaten GE foods, and 23% believed they had not (Hallman et al. 2004). Clearly, many consumers are mistaken about whether they are purchasing and consuming GE crop products.

Consumers' false beliefs probably have two sources. Many believe that companies selling foods with GE ingredients are required to indicate that fact on the label. The Rutger's Food Policy Institute (Hallman et al. 2004) found that only 33% of respondents "knew that GM foods are not required to be labeled as such in the United States," and 28% "incorrectly believed that GM foods are required to be labeled." Consumers who have no belief one way or the other about the GE status of a food prior to seeing the label might conclude from the fact that the label does not positively indicate the presence of GE ingredients that it lacks any GE ingredients. The label thus causes them to have a false belief about the product. Russo et al. (1981) call such misleadingness "incremental misleadingness."

But labels can be misleading even if they do not cause consumers to have false beliefs.[2] For example, many consumers mistakenly believe that diet margarine can be used for baking. In some cases, a label may cause this belief by some positive representation, such as a picture of a steaming muffin, and in such cases, the label would be incrementally misleading. However, consider a label that simply says "Diet Margarine" and makes no positive representation that induces consumers into believing that diet margarine can be used for baking. The label might still exhibit what Russo et al. (1981: 125) call "exploitive misleadingness" if the label exploits the fact that consumers have this mistaken belief, even though it does not cause them to hold it.[3] And, as Beales et al. (1981: 499–501) note, consumers need not even consciously hold the mistaken belief; it is enough if the existence of an implicit belief is inferable from their behavior, as evidenced by their reactions when additional corrective information is disclosed to them.

To empirically detect exploitive misleadingness, one needs to compare the beliefs of consumers who are exposed to the original label with the beliefs of consumers who are exposed to the label that has been modified by adding additional information to correct the consumer's false beliefs (Russo et al. 1981: 125–126). Presumably, some of the large number of consumers who falsely believe that unlabeled products do not contain GE ingredients would have this belief corrected by a label disclosing GE ingredients, and consumers clearly value this additional information (Streiffer and Rubel 2004). There is thus substantial research indicating that a lack of disclosure exploits consumers' mistaken beliefs about GE crop products, and that current labels are therefore exploitively misleading.

Unlike the situation with GE crop products, there are currently no GE animal products on the market, there is no consumer research measuring actual effects of GE animal product labels, and there are no data directly on the point of whether consumers will be aware of GE animal products even if they are unlabeled. Nonetheless, it is certain that a lack of labels on GE animal products will mislead *some* consumers (for the same reasons that the lack of labels on GE crops presently misleads many consumers), and this is sufficient to give the FDA and the USDA the discretion to decide whether the misleadingness is problematic enough to require labels.

## Two Objections

It is worth pausing here to address two objections. First, it might be objected that on this broad understanding of misleadingness, every label is misleading: No matter how detailed the label, there will always be some consumers who remain misinformed on things about which they care deeply. In response, it needs to be noticed that saying that information is a *candidate* for labeling is not saying that the agency *ought* to require the information on the label. Rather, the agency ought to employ some principled procedure to decide whether it should require labels. As Beales et al. (1981: 495–496) argue with respect to the Federal Trade Commission (FTC) and deceptive advertising,

> [t]he law of deception has now developed to the point of virtually eliminating any line between advertisements which are deceptive and advertisements which simply fail to inform. Indeed, it is not too broad a statement to say that present legal doctrine could make every advertisement in the country potentially deceptive. Obviously, the law has not been applied in such an extreme fashion as the FTC and the courts have stopped short of prohibiting all advertising. However, the legal definition of deception does not require any such stopping place, nor does it offer any principles to suggest where a good stopping place would be.

This point is underappreciated. The fact that misleadingness is a broad concept means that there needs to be explicit discussion of the values that should guide an agency's reaction to misleadingness, a discussion we undertake below. What matters is not only misleadingness, but also the extent to which misleadingness undermines values that are within the agency's purview. Thus, what at first looks like an objection (i.e., *all* labels are misleading) is not problematic, for it merely highlights the fact that agencies cannot rely solely on an algorithm to determine if a label is misleading, but must make judgments about values when deciding about misleadingness.

A second objection focuses on the FDCA's language that describes a label as misleading if it fails to reveal three kinds of information about the food: information relevant to the consequences of its consumption, information relevant to its suitability for customary or expected uses, and information relevant in light of positive representations made on the label (21 U.S.C. 321(n)). Fred Degnan, for example, says that the FDA's authority to require information on a label is "limited" to those three kinds of information (see chapter 3). (This is also endorsed by the FDA [Food and Drug Administration 2001].) Because the FDA believes that the process of genetic engineering is not always material in any of those ways, it does not require labeling for all GE foods (chapter 3). Thus, one might object to our analogy above between GE foods and diet margarine: Being suitable for use in baking relates to an expected use of the product, but containing GE ingredients does not.

But the FDCA prohibits misleadingness tout court, without any explicit restriction on the range of propositions about which misleadingness is problematic (21 U.S.C. § 343(a)), and section 321(n) explicitly says that those three kinds of information should be considered "among other things." Clearly, then, claiming that the FDA's authority is *limited* to requiring those kinds of information is inconsistent with the plain language of the FDCA itself (Streiffer and Rubel 2004: 228). Thus, the analogy between GE foods and diet margarine is not undermined by the distinction between being misled about facts related to expected uses and being misled about facts not so related.

## Some Potentially Relevant Differences

Of the two administrative bodies that might oversee the labeling of GE animal products, the CFSAN and the FSIS, only the FSIS has a mandatory preapproval process for labels (Food Safety and Inspection Service 2003). Before the FSIS can require or allow any new language on labels, it must institute a rule-making procedure that sets standards for the new language. This has not yet been done for GE animal products (Post 2005), so at the present time, the FSIS can neither require that GE animal products be labeled as such nor allow voluntary positive or voluntary negative labeling with respect to GE status. Until the preapproval process is completed, the FSIS cannot even allow GE animal products on the market if, in failing to disclose something that is a GE animal product, the labels are, as we argue, unacceptably misleading. That is, because the FSIS is prohibited from allowing animal products that are misleadingly labeled on the market, and would also be prohibited from requiring or allowing that they be labeled in a way that would not be misleading, the FSIS would have to prohibit the products themselves until the relevant standards had been set.

The EPIA requires that labels include any information necessary to "describe the products adequately" (21 U.S.C. §1036(a)), but the FDCA, FMIA, and PPIA do not contain such a blanket requirement. Is this relevant? Consider an example in which a meat product's label does not make any representations about the conditions under which the animal was housed, and grant, if only for the sake of the example, that such information is necessary for describing the product adequately. If consumers remain agnostic about the housing conditions, then the label does not mislead them even though they are uninformed. In such a situation, the product would be misbranded according to a policy that requires that labels disclose all material facts, but would not be misbranded according to a policy that only prohibited misleadingness. However, as we argued above, consumers do not remain agnostic in the case of GE crop products and, as we argue below, consumers are unlikely to remain agnostic in the case of GE animal products. They would instead presume that unlabeled food is not genetically engineered. In this case, then, the prohibition on misleadingness makes the requirement to disclose all material facts otiose.

The FDCA explicitly acknowledges that a lack of information can render a label misleading (21 U.S.C. 321(n)). Does the absence of such an acknowledgment in the EPIA, FMIA, and PPIA make a difference? No, because, as a matter of fact, a lack of information can render a label misleading. Since each Act prohibits misleading labels, a fortiori, each Act prohibits misleading labels where the misleadingness is due to a lack of information.

Here, we disagree with Degnan (1997: 51), who says that although the Pure Food and Drugs Act of 1906 prohibited labels that are misleading, the FDCA is stronger because, among other things, it "gives FDA the authority to require additional key information to appear on the food label if such a requirement is necessary to prevent consumers from being misled." If Degnan were correct that Acts that merely prohibit misleading labels do not also require that labels include information necessary to avoid misleadingness, then the difference between the FDCA, on the one hand, and the EPIA, FMIA, and PPIA, on the other, would mark a substantive difference. But, as mentioned, any Act that prohibits misleading labels a fortiori requires that labels include information necessary to avoid misleadingness.

### The Labeling Statutes and the Protection of Consumer Autonomy

What value or values underlie the labeling provisions of the FDCA, FMIA, EPIA, and PPIA? The case for thinking that consumer autonomy is the value underlying the labeling provisions rests on the claim that consumer autonomy is the only value capable of accounting for the breadth of those provisions. Claims made by opponents of mandatory labels that food safety, food healthfulness, and food quality exhaust the values underlying the labeling provisions are belied by several facts.

First, such claims fail to account for breadth of the prohibition on false or misleading labels present in each of the Acts: there is no requirement that the false or misleading claims have a bearing on food safety, food healthfulness, or food quality. Rather, the prohibition on false or misleading claims seems best explained by the fact that purchases made under false pretenses are not autonomous.

It would also be overly narrow to interpret the EPIA's requirement to include information necessary to "describe the product adequately" as being limited to claims relevant to food safety, food healthfulness, and food quality. For example, although the requirement is implemented in the Code of Federal Regulations as including safe handling instructions, it also includes a list of all the ingredients, added water, and the name and address of the packer or distributor (§ 590.411 (c)). These additional provisions are surely in line with the values underlying the EPIA, but they go beyond narrow considerations of food safety, healthfulness, or quality.

Second, the FDCA, FMIA, and PPIA explicitly require details that would be unjustified if food safety, healthfulness, and quality exhausted the values underlying their labeling provisions. The FDCA, FMIA, and PPIA require that imitations be labeled as such, and that labels disclose the name and place of business of the manufacturer, contain an accurate statement of quantity, disclose any artificial flavoring or coloring, list any optional ingredients for food products that meet approved standards of identity, and list all ingredients for food products with no standards of identity (21 U.S.C. § 343; 21 U.S.C. § 453 (h); 21 U.S.C. 601 (n)). (For more extensive discussion, see Streiffer and Rubel 2004; Rubel and Streiffer 2005).[4]

## CONSUMER AUTONOMY AND GE ANIMALS

The next question is whether one policy option, namely a mandatory labeling policy on GE animal products, is the only policy that adequately respects consumer autonomy, the value underlying the FDCA's labeling provisions. Our argument that it is the only such policy relies on two premises: first, the information that a GE animal product has been genetically engineered is important to consumer autonomy because it implicates their values in numerous ways; second, without mandatory positive labeling, consumers will not be adequately provided with this information.

Consumers' food purchases involving GE crop products implicate their values in numerous ways. This is evidenced by the well-documented stated preferences of many consumers that they would rather not purchase or consume GE crop products and by an understanding of the values that underlie those preferences

(Streiffer and Rubel 2004; Streiffer and Hedemann 2005). The stated preference data are also reinforced by the few revealed preference studies available (e.g., Huffman et al. 2003, Dhar and Foltz 2005).[5]

The values of consumers are even more deeply implicated in purchasing decisions about GE animal products than they are about GE crops, because consumers generally exhibit a higher degree of moral concern about animal biotechnology than about plant biotechnology. As Thomas Hoban (2002), a leading social science researcher on agricultural biotechnology, says, "Social science research over the past decade has systematically evaluated consumer perceptions of biotechnology. Results are clear that we are less likely to embrace animal biotechnology than plant-based technologies."

Hoban's summary is supported by data along two dimensions. First, research has consistently found that consumers express more concern about the moral permissibility of GE animals than about GE plants (U.S. Congress, Office of Technology Assessment 1987: 50; Pew Initiative on Food and Biotechnology 2003a ; Hoban 2003; Hoban and Kendall 1993: 13; Hoban et al. 1992: 482). Second, research has consistently found that consumers express a significantly greater aversion to the idea of eating GE animal products than to the idea of eating GE plant products (Hallman et al. 2002: 20–22; Hallman et al. 2003: 10–11; Hallman et al. 2004: 7–8). These findings are summarized in table 5.1.

These findings are easily explicable in terms of consumers' values: Many consumers have moral concerns about genetic engineering in general (U.S. Congress, Office of Technology Assessment 1987; Streiffer and Rubel 2004; Streiffer and Hedemann 2005); additionally, they perceive GE animals as raising numerous ethical issues over and above those raised by GE plants. The fact that these poll results reflect consumers' values is crucial in determining whether information about GE status is necessary for consumers to exercise their autonomy in their food purchases.

There are a numerous consumers' values that make information about the GE status of animal products important, including concerns about animal welfare, the inappropriate genetic alteration of animals, slippery slopes toward human genetic engineering, risks to the environment, and food wholesomeness. We address these in turn.

Consumers clearly have a direct concern for the welfare of GE animals, and a concomitant reluctance to support or participate in practices they view as detrimental to animal welfare. Indeed, once the phrase "animal rights" is properly distinguished from absolutist views that would prohibit all animal use (Varner 2002), the vast majority of Americans believe that animals have rights that protect animal welfare in important ways (Hoban and Kendall 1993; Craig and Swanson 1994; Mench 1999). Bernard Rollin (1995: 179; see also 143–169) has argued that there is an emerging social consensus in support of "a clear social maxim" that requires that GE animals "be no worse off, in terms of suffering, after the new traits are introduced into the genome than the parent stock was prior to the insertion of the new genetic material."

If Rollin is correct, and it does seem borne out by empirical studies (see Mench 1999, 253–255), then the National Research Council (2004) report on animal biotechnology does not bode well for expecting GE animal products to be

**Table 5.1. Consumer Preferences about GE Plants Compared to Those about GE Animals**

| Year | Source | Plants | Animals |
|---|---|---|---|
| 2004 | Food Policy Institute, Rutgers (Hallman et al. (2004) | 41% "leaned toward disapproval or disapproved" of plant-based GE food | 61% "leaned toward disapproval or disapproved" of animal-based GE food |
| 2004 | Hossain and Onyango (2004) | 24% were "completely unwilling" to consume GE orange juice with a transgene from a plant | 44% were completely unwilling if transgene was from an animal |
| 2004 | Hossain and Onyango (2004) | 36% were "completely unwilling" to consume GE beef hamburger with a transgene from a plant | 40% were "completely unwilling" if the transgene was from an animal |
| 2003 | The Mellman Group, for the Pew Intiative on Food and Biotechnology (2003a) | 6.08 (out of 10) mean comfort level with GM plants | 3.81 (out of 10) mean comfort level with GM animals |
| 2003 | The Mellman Group, for the Pew Initiative on Food and Biotechnology (2003a) | 81% said that "producing more affordable pharmaceuticals was a good reason to genetically modify plants"; 14% said it was a bad reason | 49% said that it was a good reason; 41% said it was a bad reason |
| 2003 | Food Policy Institute, Rutgers (Hallman et al. 2003) | 39% "leaned toward disapproval or disapproved" of plant-based GE food | 66% "leaned toward disapproval or disapproved" of animal-based GE food |
| 2002 | Food Policy Institute, Rutgers (Hallman et al. 2002) | 37% disapproved strongly or somewhat of GE plants; 22% said it was "morally wrong" | 68% disapproved strongly or somewhat of GE animals; 55% said it was "morally wrong" |
| 1993 | Hoban and Kendall (1993) | 24% think that it is "morally wrong to genetically engineer plants" | 53% think it is "morally wrong to genetically engineer animals" |
| 1992 | Hoban et al. (1992) | 23% "oppose plant genetic engineering" | 53% "oppose animal genetic engineering" |
| 1987 | U.S. Congress, Office of Technology Assessment (1987) | 6.6 (out of 10) mean acceptability level with GE plants | 5.3 (out of 10) mean acceptability level with GE animals |

consistent with consumers' values; as we summarize below, it identifies numerous ways in which genetic engineering increases risks to animal welfare.

Genetically engineered animals produced through in vitro embryo culture techniques frequently suffer from large-offspring syndrome, which causes higher birth weights, longer gestational periods, more congenital malformations, and higher perinatal mortality rights (National Research Council 2004: 95). And even when the animals do not suffer from large-offspring syndrome, they still "seem to be less viable and more often experience problems like double-muscling, leg and joint problems, hydroallantois, heart failure, enlarged organs, and cerebellar dysplasia" (National Research Council 2004: 95). Although GE animals will not transmit these problems to their offspring, which are not themselves produced through in vitro embryo culture, the offspring are still at greater risk for inheriting retarded growth and abnormal DNA methylation patterns from their parents (National Research Council 2004: 96).

Although the technology continues to improve, problems persist with variability of transgene expression. Problems are most widely documented with growth hormone genes, the genes most likely to be utilized in GE animals intended for food, and include increased risk of "diarrhea, mammary development in males, lethargy, arthritis, lameness, skin and eye problems, loss of libido, and disruption of estrous cycles," as well as "increased susceptibility to stress" (National Research Council 2004: 98). Growth hormone transgenes in salmon increase risk of "cranial deformities and opercula overgrowth" (National Research Council 2004: 98).

The insertion of the transgene can interrupt the function of other genes, with a frequency that the National Research Council (2004) estimates at about 5–10%. Most of the mutated animals never reach term, but those that do experience problems such as "severe muscle weakness, missing kidneys, seizures, behavioral changes, sterility, disruptions of brain structure, neuronal degeneration, inner ear deformities, and limb deformities" (National Research Council 2004: 97).

Animals genetically engineered to produce pharmaceuticals may experience physiological effects from the substances, and those that are engineered to produce pharmaceuticals in their milk may experience painful lactation (National Research Council 2004: 102). Because of their high commercial value, they are likely to receive exceptional veterinary care; however, they are also likely to be weaned early and kept in barren, easily sanitized environments, which are likely to be psychologically stressful (Reiss and Straughan 1996). The same is true of GE animals produced as sources of organs for xenotransplantation (National Research Council 2004: 103).

Subsequent generations of GE animals reproduced through cloning (which ensures the transmission of desired genotypes) also face an array of welfare problems arising from the cloning process itself (National Research Council 2004: 100–102).

In addition to direct harm to animal welfare, there may also be systemic effects resulting from the use of GE a nimals. The National Research Council (2004) observes that even something as apparently beneficial as genetically engineering an animal for disease resistance can have a negative overall impact on animal welfare if it encourages producers to keep the animals even more closely confined.

Moreover, many consumers think it is inappropriate to use genetic engineering to alter animals so that they suffer less when the suffering is due to the conditions under which they are kept (Mench 1999: 254). Because animal suffering is a result of multiple factors deriving both from the animals' physiology and the production environment, it will sometimes be possible to relieve suffering either by altering the animal or by altering the production environment But where animals suffer because of production practices, many consumers resist using genetic engineering to make the animals better able to endure those practices, and they reasonably view such engineering as an abdication of producers' responsibility to provide production conditions suitable for the animals.[6]

Other considerations also implicate consumers' values. Hoban (2002) notes that people have greater concern for animals because of their emotional bonds with their pets. Moreover, the perception that animals are evolutionarily closer to humans raises slippery slope concerns about genetically engineering humans, concerns that certainly appear vindicated by the history of biotechnology development. The greater mobility of animals raises greater environmental worries in some people's minds and has received intense discussion in the case of GE salmon, as we discuss below. And although there is broad public support for using GE animals as a source of humanized organs for xenotransplantation, "the thought of consuming meat from such animals is unsavory and unacceptable to the vast majority of people" (Hoban 2002).

It is of course true that many of the ethical concerns consumers have about GE animal products are not unique to genetic engineering, and it is also true that GE animals are not necessarily worse off than or more problematic than some of their conventional counterparts. But neither of these needs to be true in order for mandatory labeling to be justified. For example, the FSIS, under the PPIA, requires that any raw poultry product that has ever been frozen must be labeled as "frozen" (98 C.F.R. 381.129 (6)(ii)), even though freezing is not the only thing that can damage product quality, nor is frozen poultry necessarily worse than nonfrozen poultry. The point is that genetic engineering, as it is likely to be implemented, will conflict with many consumers' values, especially the value of animal welfare. To a large extent, genetic engineering is going to be directed by profit-motivated actors and, hence, by those whose interests lie in intensification and increased productivity. These interests are, for most practical purposes, at odds with those of the animals, and so there is a presumption that any technology that expands the range of alterations capable is likely to result in greater suffering of the animals. As Rollin (1995: 181) notes,

> [O]pponents of genetic engineering of animals are right to fear that such engineering will proliferate animal suffering, though they are wrong to think that it must do so. In other words, if genetic engineering is allowed to proceed unconstrained, it is likely that concerns for animal welfare, animal suffering, and animal happiness will occupy positions of low priority as compared with considerations of profit and short-run "efficiency."

More generally, consumers are quite reasonably concerned to make their food purchasing decisions in light of their awareness of the practices they are supporting with their money. Thus, respecting consumer autonomy requires that people have that information available to them so that they can act on it as they deem appropriate.

## CONSUMER AUTONOMY AND MANDATORY POSITIVE LABELS

So, the information that a GE animal product has been genetically engineered is important to consumer autonomy. The next question in determining whether mandatory labeling is the only policy option that adequately respects consumer autonomy is whether mandatory positive labels are necessary for adequately informing consumers. The situation to date makes it look unlikely that consumers will be informed without mandatory labels. As mentioned above, the existing lack of positive labels on GE crop products has been and continues to be demonstrably misleading to many consumers, despite the presence of off-label information and a robust, voluntary, negative labeling alternative, namely, organics. Thus, off-label methods of providing information about GE status have been ineffective, and negative labeling is not sufficient to prevent those who buy nonorganic foods from being misled by the lack of information (Rubel and Streiffer 2005). The final alternative would be voluntary positive labeling, which industry has refused to adopt. Are there any differences between GE crop products and GE animal products that would give grounds for a more optimistic prediction in the case of GE animals?

There are several potentially relevant differences. Part of the reason that so many consumers are uninformed about the prevalence of GE crop ingredients is that GE crop ingredients became prevalent in the U.S. food supply very quickly. Roundup Ready soybeans and Bt corn were first grown commercially in 1996 but constituted 85% and 45% of the 2004 U.S. soybean and corn markets, respectively (Pew Initiative on Food and Biotechnology 2004b). It seems unlikely that GE animals would become so prevalent so quickly. Technical obstacles, such as the need for multigenerational studies in breeds with long reproductive intervals (Pursel 1989), will delay the development of new GE animals, and consumer acceptance will undoubtedly require "a long and difficult educational process" (Hoban 2002). It is also likely that there will be a great deal of attention paid to the first GE animal products on the market. Both the Enviropig and the AquAdvantage Bred salmon have already received some mainstream media coverage.

However, none of these considerations provides good grounds for thinking that the case of GE animal products will significantly differ from GE crop products with respect to how well informed consumers are as to their prevalence. As GE animal products begin to filter through the U.S. food supply, their novelty will likely mean that few consumers are aware of them, even if they are introduced with a great deal of fanfare: Significant fanfare about the Flavr-Savr tomato, the first GE whole-food product on the U.S. market, failed to promote significant awareness of GE crop products. Moreover, the public's current lack of awareness of the prevalence of GE plant products persists despite the fact that those foods have been on the market for years, and despite the existence of publicly available information about those foods being on the market.[7] Consumer acceptance of GE animal products probably will require a long and difficult educational process, as Hoban (2002) predicts, but if they are unlabeled, consumers may purchase them long before they are accepted.

Two conclusions, therefore, seem warranted. First, it seems most likely that despite some differences between GE animal products and GE crop products, the situation with GE crops will be largely replicated in the case of GE animals, with most consumers being unaware of whether the foods they are consuming contain GE animal products. Second, once GE animal products are on the market, it is incumbent on those who would argue against mandatory labels to perform the relevant research to show that they are not necessary for respecting consumer autonomy. Until that research is done and we have direct data on the issue, respecting consumer autonomy requires mandatory positive labeling on all GE animal products.

It might be objected that mandatory positive labels on GE foods will themselves undermine autonomy by misleading consumers (e.g., because they will be misperceived as a food safety warning). But from the perspective of consumer autonomy there is an important asymmetry between providing true information from which consumers will draw mistaken conclusions and withholding true information that will leave consumers with mistaken beliefs. The former still leaves the decision in the hands of the consumer, to decide in accordance with his or her values as he or she sees fit. Thus, from the perspective of autonomy, labels that mislead by failing to disclose information are more problematic, other things equal, than labels that mislead by disclosing information. There is also a second asymmetry between a misleading positive label and a label that misleads due to failure to disclose, namely, that a positive label can prompt consumers to look for further information whereas a failure to disclose does not.

## ARE THERE SUFFICIENTLY STRONG COUNTERVAILING REASONS FOR DEVIATING FROM THE MOST DEMOCRATICALLY LEGITIMATE POLICY?

The final question is whether there are sufficiently strong countervailing reasons that would justify the FDA or the USDA deviating from applying the labeling statutes in accordance with their underlying values. Such reasons would include gross unfairness, a lack of democratic legitimacy in the procedures leading to the decision, prohibitive economic costs, and paternalistic considerations. We have argued elsewhere that these do not obtain with respect to mandatory positive labeling on GE crop products (Streiffer and Rubel 2004: 233–234, 236–242), and we see no reason at this time to suspect that the situation will be significantly different with respect to GE animal products.[8]

## THE CVM, ACTING UNDER THE LABELING PROVISIONS OF THE ANIMAL DRUG PROVISIONS OF THE FDCA

At the present time, the FDA has asserted that it will regulate GE animals under the new animal drugs provisions of the FDCA (Pew Initiative on Food and Biotechnology 2004a: 7; Food and Drug Administration, Center for Veterinary Medicine n.d.). The CVM is authorized to require labeling for new animal drugs (21 U.S.C. §352), and it seems clear that these labeling provisions have two justifications: first, to protect the autonomy of the purchasers, in this case, veterinarians or farmers; and second, to provide information necessary to ensure that those administering the drug do so in the proper way. However, it is implausible to suppose that these labeling provisions would make sense applied to labels intended for consumers in the grocery store since they are instead intended to be used by the veterinarians administering and overseeing the application of drugs on the farm. (For more discussion, see Pew Initiative on Food and Biotechnology 2004a: 120).

## SUMMARY

Thus, each of these different regulatory pathways will support a mandatory positive labeling policy, because each statute shares features that, first, provide the regulating agency discretion to require labels and, second, provide a democratically based reason to use that discretion to require labels. Where an agency has the discretion to act (in this case, to require labels), the decision about how to exercise that discretion ought to be governed by the underlying justifications for the statutes that grant discretion in the first place. Because the underlying justification for the labeling provisions of the FDCA, FMIA, EPIA, and PPIA is the protection of consumer autonomy, and because requiring positive labels on GE animal products is the only policy that adequately respects consumer autonomy, these agencies ought to require mandatory positive labels on GE animal products under their purview. The relevant details of the regulatory pathways are summarized in table 5.2.

## EXAMPLES

Up to this point we have focused in broad terms on the administrative agencies' authority to require labels on GE animal products. In this section, we examine the particular issues raised by the two products nearest commercialization. The first, the AquAdvantage Bred Salmon, appears to confirm the appropriateness of the framework argued for above. The second, the Enviropig, is different in important ways from other GE foods and from the transgenic salmon. Considered in isolation, those differences arguably provide a reason against mandatory labeling; however, within the context of establishing labeling policy appropriate for a potentially wide array of GE animal products, those differences do not justify eschewing a mandatory labeling scheme.

### AquAdvantage Bred Salmon

AquaBounty, in conjunction with A/F Protein, has developed GE salmon by inserting a growth hormone gene that allows the salmon to reach market size about twice as fast as conventional salmon (Fletcher et al. 2000; Union of Concerned Scientists 2001; Martin 2003). The salmon are trademarked as "AquAdvantage Bred salmon," and their shortened harvest time could prove advantageous to many aquaculture operations, but was primarily developed so that the salmon could be

**Table 5.2.** Summary of Regulatory Pathways

| | | Does the Agency Have Discretion? | | | Does Autonomy Underlie the Provisions? | | | |
|---|---|---|---|---|---|---|---|---|
| Agency | Provisions | Prohibits False Claim | Prohibits Misleading Claims | Expressly Considers Lack of Information | Requires All Material Facts | Prohibits False Claim | Prohibits Misleading Claims | Specifically Requires Other Information |
| FDA-CFSAN | FDCA; 21 U.S.C. §§ 321, 343 | Y | Y | Y | N | Y | Y | Y |
| USDA-FSIS | FMIA; 21 U.S.C. § 601 | Y | Y | N | N | Y | Y | Y |
| USDA-FSIS | EPIA; 21 U.S.C. § 1031 | Y | Y | Y | Y | Y | Y | N |
| USDA-FSIS | PPIA; 21 U.S.C. § 451 | Y | Y | N | N | Y | Y | Y |
| FDA-CVM | FDCA; 21 U.S.C. § 301 | Y | Y | N | N | Y | Y | Y |

Abbreviations: CFSAN, Center for Food Safety and Applied Nutrition; CVM, Center for Veterinary Medicine; EPIA, Egg Products Inspection Act of 1970; FDA, U.S. Food and Drug Administration; FDCA, Food, Drug, and Cosmetic Act of 1938; FMIA, Federal Meat Inspection Act of 1906; FSIS, Food Safety and Inspection Service; PPIA, Poultry Products Inspection Act of 1957; USDA, U.S. Department of Agriculture.

farmed without overwintering, opening up locations for aquafarms previously inaccessible due to freezing winter temperatures (Du et al. 1992; Fletcher et al. 2000).

The AquAdvantage salmon case fits our model well for several reasons. To begin, in the absence of a label, many consumers will be misled into thinking that the AquAdvantage is a conventional fish. Second, it is unlikely that AquAdvantage producers will voluntarily label the fish as GE: There is no reason to suspect that consumers would overall prefer GE fish (and substantial reason to suspect they will be averse to them [Hoban 2003]), and there is therefore no incentive to voluntarily label. Third, and relatedly, the aspects of the fish about which consumers will be misled implicate values that matter deeply to consumers. Being misled about such matters undermines the autonomy of their purchasing decisions.

The two primary values implicated by the AquAdvantage salmon are environmental protection and animal welfare. The environmental risk of AquAdvantage salmon is premised on their ability to escape, survive, and reproduce outside of captivity. The high likelihood and frequency of escape for farm-raised fish (sometimes on the order of hundreds of thousands in a single year) is well documented (Kelso 2004; National Research Council 2004: 90), and it therefore seems likely that some AquAdvantage salmon will escape.

If this occurs, there are three possibilities for environmental impact. First, the fish might escape and reproduce, but the transgene might be quickly eliminated from the wild population. Second, the transgenic fish might become stably incorporated into the wild population. Third, the transgenic fish might initially become incorporated into the wild population and subsequently cause it to crash.

The first of these does not seem to be a significant environmental problem; the second is only arguably so, depending on whether introducing a GE organism is ipso facto an environmental harm or whether it must have some harmful effect beyond its mere introduction. The third, though, is a substantial problem on any plausible view, and there is reason to think that it would occur. Howard Muir and Richard Howard (1999) argue that fish that grow faster because they have been engineered with growth hormone (like the AquAdvantage) have a substantial risk of exhibiting a "Trojan gene" effect. That is, they have better mating success, which allows the transgene to spread through the population, but generate offspring with reduced viability. This is environmentally problematic because the GE fish decrease the numbers of their conventional counterparts while replacing them with less fit individuals. Over time, this process leads to decreasing viability and eventually to the extinction of the entire population.

There are two possible ways to avoid the third situation. One can restrict the use of growth-enhanced salmon to inland fisheries, or one can induce sterility in the salmon. But to say that there are ways of managing the environmental risk of growth-enhanced salmon is a far cry from ensuring the risk will actually be managed, and given the current gaps in the U.S. regulatory structure, it is not clear that the overseeing agency will have the authority to place geographic restrictions in the name of managing environmental risk. Further, techniques for inducing sterility are never 100% effective (Kelso 2004: 516), and the models developed by Muir and Howard (1999) indicate that population crashes can occur even with very few original escapees.

The other value implicated in consumer purchases of the AquAdvantage salmon is animal welfare. The National Research Council (2004) identifies numerous ways in which growth-hormone–enhanced salmon do worse in terms of animal welfare than their conventional counterparts. For example, the growth hormone transgene increases the risks of growth abnormalities in the head and jaw, which can result in eating difficulties and starvation, and these deformities are passed on to the GE salmon's progeny, and even become more pronounced (National Research Council 2004: 98–99). Growth hormone transgenes can also have adverse effects on smoltification, gill irrigation, disease resistance, body morphometry, pituitary gland structure, life span, larval development rate, swimming ability, feeding rates, and risk avoidance behavior (National Research Council 2004:99–100).

Consumers who care about whether their purchases support or endorse practices that harm the environment or animal welfare will be justified in their concern about the AquAdvantage Bred salmon. If those salmon are not labeled, many such consumers will purchase the AquAdvantage Bred salmon when they would not have if they had been informed (and others would at least like to have that information available to them, even if it might not affect their purchasing behavior). Thus, a lack of a mandatory labeling policy seems likely to undermine the autonomy of consumers. For the reasons argued above, this means that the relevant agency has a democratic obligation, grounded in its obligation to use its discretion in ways that further the underlying rationale for the labeling statutes, to mandate labels.

## Enviropigs

The Enviropig provides an interesting test case for GE animal product labeling because some of the details surrounding it do not match the general picture of GE animal products we outline above. While we think that schema is justified, the Enviropig presents some illuminating problems for it.

Pigs have a dietary requirement for phosphorus, but standard plant-based feed contains only plant phytate, which pigs are unable to digest and which therefore becomes concentrated in their manure. The most common way to provide pigs with sufficient amounts of phosphorus has been to add supplemental phosphorus to the pigs' feed in the form of mineral phosphate. However, supplemental phosphate has the unfortunate side effect of increasing the amount of phosphorus in the pigs' manure. Because the manure is used as fertilizer, this means that supplemental phosphate has the side effect of increasing phosphorus in the environment. Phosphorus pollution from the manure of intensive hog farms is an enormous environmental problem (Jongbloed and Lenis 1998; Carpenter et al. 1998), and so farmers seek ways to mitigate it. Prior to the development of the Enviropig, the most effective mitigation technique was to add phytase to standard feed. The phytase allows pigs to digest the plant phytate, getting the phosphorus they need, and reduces the amount of phosphorus secreted by nontransgenic pigs by up to 56%.

Researchers at the University of Guelph, funded by Ontario Pork, genetically engineered pigs to produce phytase in their salivary glands (Forsberg et al. 2003).

Enviropigs fed standard feed (even without supplemental phytase) do even better than conventional pigs fed feed supplemented with phytase, and secret up to 75% less phosphorus (Golovan et al. 2001; Harper et al. 2003). These pigs are touted as an environmentally friendly application of genetic engineering and, hence, have been dubbed "Enviropigs" by their marketers.

Given these facts, there are four features of the Enviropigs that might be touted as selling points to consumers, three of which are alleged to follow from the first:

1. Enviropigs produce less phosphorus in their manure per pig than do conventional pigs using even the most effective alternative methods of phosphorus reduction.
2. Hog farms using Enviropigs will produce less overall phosphorus pollution than will other hog farms, even those using the most effective alternative methods (Forsberg et al. 2003: E68).
3. Enviropigs will benefit farmers in developing countries because raising Enviropigs is more affordable than adding phytase to their feed or purchasing low-phytate feed (Forsberg et al. n.d.).
4. Because Enviropigs can extract more useful phosphorus from their feed, the use of Enviropigs will reduce strain on the "dwindling source of economically recoverable mineral P [phosphorus]" (Forsberg et al. 2003: E72; Smil 2000).

These alleged benefits provide a reason for voluntary labeling because consumers are willing to pay more for products that they perceive as having environmental benefits. Similarly, some studies indicate that consumers wary of GE products are more willing to purchase them if the product is engineered to have an environmental benefit (Forsberg et al. 2003; Bruhn 2003). Thus, adverting to the Enviropig's features will create incentives for consumers to purchase food derived from the Enviropig and, moreover, may offset consumer aversion to GE foods. Thus, at least in cases such as the Enviropig, it might be argued that these incentives are strong enough to obviate the need for mandatory labels. In the Enviropig case, it is easy to imagine a voluntary labeling scheme, where the primary label says "Environmentally Friendly," and a longer description touts features 1–4 on a secondary label display.

There are two problems with this argument against the need for mandatory labels. First, it presupposes that Enviropigs will have the consequence of reducing overall environmental pollution. But right now, as Clare Schlegel, chairman of Ontario Pork, says, "The environmental barriers are the largest in terms of growing as an industry" (quoted in Vestel 2001). Thus, even if each Enviropig will produce less phosphorus, because phosphorus pollution is the limiting factor to the size of many intensive hog operations, using Enviropigs could result in an increase in the number of pigs, with the result that the overall phosphorus pollution will remain the same, and other pollutants will increase (*Minnesota Daily* 2001). That is, using Enviropigs will engender precisely the sort of agricultural intensification that gives rise to many consumers' concerns about modern agriculture and other GE products.

Moreover, even if the critics are wrong, and features 2–4 really do obtain, there remains a disincentive for the Enviropig label to explicitly mention GE, as opposed to merely mentioning its environmental benefits. In cases of value tradeoffs (here between aversion to GE and interest in environmentally friendly products), it best respects autonomy to provide both sets of facts to consumers and let them decide for themselves how to make the tradeoff.

And finally, even if the case for mandatory labeling is weak in the case of Enviropigs, this undermines the case for a mandatory labeling requirement only if they either are the only GE animal product on the market or are indicative or representative of GE animal products in general. As more products are developed and reach the market, neither of these seems likely to be true.

## CONCLUSION

The primary justification for labeling provisions is not the need to alert consumers to food health or safety issues. Rather, labeling provisions are better understood as a mechanism for protecting consumers' autonomy by providing them with information important to their purchasing decisions. Given the wide range of important values implicated by animal biotechnology, information about GE animal products will be vital for many consumers' decisions about what animal products to purchase. We argue that they will not have that information absent mandatory positive labeling on GE animal products. This is especially clear in the case of AquAdvantage Bred salmon. With respect to Enviropigs, the case is more ambiguous; nonetheless, we think that upon close examination, regulatory agencies ought to exercise their discretion to require labels on products derived from Enviropigs, too.

Given the importance of consumers' values in this debate, and given the dearth of direct empirical research about consumers' views about GE animals, GE animal products, and labeling on GE animal products, we think it is especially important for social scientists to pursue research about those values, on par with the extensive research on the labeling of GE crop products. This should happen both before the products come to market and as they become a larger part of the food supply.

## Notes

For their many helpful comments, we thank Molly Gardner and the participants of the Labeling of Genetically Modified Food Conference at the University of Missouri–Columbia, especially Peter Markie, who gave extensive and careful consideration to our earlier arguments about labeling.

1. We are puzzled by Peter Markie's interpretation in chapter 6 of our arguments in Streiffer and Rubel (2004) and Rubel and Streiffer (2005) as addressing marketers of genetically modified (GM) products; we explicitly restricted our attention to the FDA, Congress, or, most broadly, public policy.
2. Here, we would disagree with Markie's view that marketers discharge their obligation not to deceive their customers as long as they "are not responsible for the creation of consumers' false belief in the first place" (chapter 6).
3. Beales et al. (1981) discuss several other relevant examples, including a used product that does not disclose that it is used. That would be misleading even though a used product need not be physically different or inferior from a new product. And this can be misleading even

if there is no positive representation made on the label regarding whether the product is new or used.

4. Markie (chapter 6) notes that in Streiffer and Rubel (2004), we did not rule out several other potential explanations, including the promotion of fairness as the rationale for requiring an accurate quantity statement and the prevention of deception for its own sake as the rationale for prohibiting false or misleading claims. We are dubious that fairness provides a defensible rationale: Should the FDA allow a company to mislabel a product's quantity as long as it charges a price that is fair given its true quantity? No, because although such a transaction would be fair, it would not be autonomous. And it is hard to make sense of the idea that deception is to be avoided for its own sake, when deception seems to be problematic only when it undermines autonomy. Deception at the magic show, for example, is not something to be avoided for its own sake. At any rate, in order for Markie's alternative rationales to undermine our conclusion, he also needs to show that they would not support a mandatory labeling policy on GE foods. He does not address this in the case of fairness, and his only argument as to why the prevention of deception does not justify a mandatory labeling policy is that the current labels do not cause consumers' false beliefs and so are not deceptive. But as we argue below, there is evidence that existing labels do cause consumers' false beliefs in some cases, and a label can be misleading, and hence, deceptive, in other ways.

5. Kalaitzandonakes et al. (chapter 7) discuss the data that the presence of GE labels did not significantly affect aggregate purchasing behavior of GE soy products in the Netherlands or China, and they suggest that this undermines the case for labels based on consumer autonomy. From the perspective of consumer autonomy, such revealed preference data are important to consider, but Kalaitzandonakes et al. oversimplify the relationship between (1) whether GE labels affect aggregate purchasing behavior and (2) whether information about GE labels is important to consumer autonomy. First, information that alters behavior is not always crucial to autonomy: A house seller's failure to disclose his reservation price to a buyer does not undermine the autonomy of the buyer's decision, even if it would have affected the transaction price. Second, and more important, information does not have to affect behavior in order for its provision to be required by autonomy: Having a medical student practice a procedure on a patient without disclosing the student's status violates the patient's autonomy even if the patient would have consented if informed. The FDA's focus group work is relevant here (Food and Drug Administration 2000): Upon being told about the prevalence of GE foods in the U.S. food supply, "[t]he typical reaction of participants was not one of great concern about the immediate health and safety effects of unknowingly eating bioengineered foods, but rather outrage that such a change in the food supply could happen *without them knowing about it*" (emphasis added). Moreover, in populations where negative attitudes toward GE foods are roughly as prevalent as positive attitudes, labels could make a significant difference at the individual level without much effect on aggregate behavior.

6. This contrasts with the view expressed by William Muir (quoted in Sigurdson 1995), who is quoted as saying that given a choice between altering the animals and altering the production practices in order to relieve suffering, altering the animals sometimes "makes more sense."

7. A free website maintained by Cornell extension, for example, summarizes in lay language how a consumer can predict if a particular product is likely to contain GE ingredients (Genetically Engineered Organisms-Public Issues Education Project 2004).

8. Markie (chapter 6) interprets our consumer autonomy and democratic equality arguments (Streiffer and Rubel 2004) as assuming that "consumer autonomy outweighs any of the other considerations we regularly take to be relevant to a labeling decision, no matter how strong those other considerations may be." Similarly, Kalaitzandonakes et al. (chapter 7) claim that consumer autonomy arguments "sidestep key issues such as the relevance of other instruments

for avoidance of market failure (e.g., voluntary labeling) or the size of compliance costs and the concomitant efficiency of mandatory labels." In fact, the arguments assume no such thing. We agree that there could be sufficiently strong countervailing considerations both for the FDA and for Congress but have argued at length that they do not obtain in the case of mandatory GE labeling (Streiffer and Rubel 2004, 233–234, 236–242) and that voluntary labeling is not up to the task (Rubel and Streiffer 2005). The only one of those arguments that Markie takes up in chapter 6 is our claim that a mandatory labeling policy is not unfair to purveyors of GE foods merely in virtue of reducing their profits even though their products are the same in terms of safety, quality, and healthfulness. He points out, rightly, that a car dealer who merely fails to disclose an upcoming sale that would have affected a consumer's purchase is nonetheless entitled to the profits he makes if the sale goes through. But the issue is not whether sellers are entitled to their past profits given that they complied with existing regulations; they probably are in most cases. The question is whether a change in the regulations would be unfair solely because it would reduce the profits of some by making consumers more informed about their products even though those products are the same as others in terms of safety, quality, and healthfulness. Given that we were not, of course, arguing that the FDA or Congress should take back some percentage of past GE food sales and return it to consumers, Markie's example fails to be probative.

## References

Animal and Plant Health Inspection Service. n.d. United States Regulatory Oversight in Biotechnology—Responsible Agency Overview. Retrieved from www.aphis.usda.gov/brs/usregs.html#usda. Accessed June 17, 2005.

Beales, H., Craswell, R., and Salop, S. C. 1981. The Efficient Regulation of Consumer Information. *Journal of Law and Economics* 24: 491–539.

Belson, N. 2000. US Regulation of Agricultural Biotechnology: An Overview. *AgBioForum* 3(4): 268–280.

Bruhn, C. M. 2003. Consumer Attitudes toward Biotechnology: Lessons from Animal-Related Applications. *Journal of Animal Science* 81 (e. suppl. 2): E196–E200.

Carpenter, S., Caraco, N. F., Correll, D. L., Howarth, R. W., Sharpely, A. N., and Smith, V. H. 1998. Nonpoint Pollution of Surface Waters with Phosphorus and Nitrogen. *Issues in Ecology* (3).

Craig, J. V., and Swanson, J. C. 1994. Review: Welfare Perspectives on Hens Kept for Egg Production. *Poultry Science* 73: 921–938.

Degnan, F. 1997. The Food Label and the Right-to-Know. *Food and Drug Law Journal* 52: 49–60.

Dhar, T., and Foltz, J. 2005. Milk by Any Other Name…Consumer Benefits from Labeled Milk. *American Journal of Agricultural Economics* 87(1): 214–228.

Du, S. J., Gong, Z., Fletcher, G. L., Shears, M. A., King, M. J., Idler, D. R., and Hew, C. L. 1992. Growth Enhancement in Transgenic Atlantic Salmon by the Use of an "All Fish" Chimeric Growth Hormone Gene Construct. *Bio/Technology* 10 (February): 176–181.

Egg Products Inspection Act of 1970. U.S. Public Law 597. 91st Cong., 2d Sess., 29 December 1970.

Federal Meat Inspection Act of 1906. U.S. Public Law 242. 59th Cong.

Fletcher, G. L., Goddard, S. V., and Hew, C. L. 2000. Current Status of Transgenic Atlantic Salmon for Aquaculture. In *Proceedings of the 6th International Symposium on The Biosafety of Genetically Modified Organisms*, ed. C. Fairbairn, G. Scoles, and A. McHughen. University Extension Press, Canada. Retrieved from http://www.ag.usask.ca/isbr/Symposium/Proceedings/Section9.htm. Accessed June 17, 2005.

Food and Drug Administration. 2000. Report on Consumer Focus Groups on Biotechnology. Retrieved from www.cfsan.fda.gov/~comm/biorpt.html. Accessed October 28, 2006.

Food and Drug Administration. January 18, 2001. Draft Guidance for Industry, Voluntary Labeling Indicating Whether Foods Have or Have Not Been Developed Using Bioengineering. *Federal Register* 66(12): 4839–4840.

Food and Drug Administration. 2003. FDA Investigated Improper Disposal of Bioengineered Pigs. Retrieved from www.fda.gov/bbs/topics/ANSWERS/2003/ANS01197.html. Accessed June 17, 2005.

Food and Drug Administration, Center for Veterinary Medicine. n.d. Questions and Answers about Transgenic Fish. Retrieved from www.fda.gov/cvm/transgen.htm. Accessed November 18, 2005.

Food, Drug, and Cosmetic Act of 1938. U.S. Public Law 717. 75th Cong., 3d Sess., 25 June 1838.

Food Safety and Inspection Service. 2003. Labeling of FSIS Regulated Foods. Retrieved from www.fsis.usda.gov/OPPDE/larc/Policies/Label101/Label101.PPT. Accessed June 17, 2005.

Forsberg, C., Phillips, J. P., Golovan, S. P., Fan, M. Z., Meidinger, R. G., Ajakaiye, A., Hilborn, D., and Hacker, R. R. 2003. The Enviropig Physiology, Performance, and Contribution to Nutrient Management Advances in a Regulated Environment: The Leading Edge of Change in the Pork Industry. *Journal of Animal Science* 81 (14 suppl. 2): E68–E77.

Forsberg, C., Golovan, S., Ajakaiye, A., Phillips, J. P., Meidinger, R. G., Fan, M. Z., Kelly, J. M., and Hacker, R. R. n.d. Genetic Opportunities to Enhance Sustainability of Pork Production in Developing Countries: A Model for Food Animals [Slide Presentation]. Retrieved from www.iaea.org/programmes/nafa/d3/public/forsberg-fao-iaea-web.pdf. Accessed June 15, 2005.

Genetically Engineered Organisms-Public Issues Education Project. 2004. GE Foods in the Market. Retrieved from www.geo-pie.cornell.edu/crops/eating.html. Accessed June 17, 2005.

Golovan, S. P., Meidinger, R. G., Ajakaiye, A., Cottrill, M., Wiederkehr, M. Z., Barney, D. J., Plante, C., Pollard, J. W., Fan, M. Z., Hayes, M. A., Laursen, J., Hjorth, J. P., Hacker, R. R., Phillips, J. P., and Forsberg, C. W. 2001. Pigs Expressing Salivary Phytase Produce Low-Phosphorus Manure. *Nature Biotechnology* 19(8): 741–745.

Hallman, W. K., Adelaja, A. O., Schilling, B., and Lang, J. T. 2002. Public Perceptions of Genetically Modified Food: Americans Know Not What They Eat. Report No. RR-0302-001. New Brunswick, N.J.: Food Policy Institute, Cook College, Rutgers University.

Hallman, W., Hebden, W. C., Aquino, H. L., Cuite, C. L., and Lang, J. T. 2003. Public Perceptions of Genetically Modified Food: A National Study of American Knowledge and Opinion. Report No. RR-1104-0007. New Brunswick, N.J.: Food Policy Institute, Cook College, Rutgers University.

Hallman, W., Hebden, W. C., Aquino, H. L., Cuite, C. L., and Lang, J. T. 2004. Americans and GM Food: Knowledge, Opinion, and Interest in 2004. Report No. RR-1104-0007. New Brunswick, N.J.: Food Policy Institute, Cook College, Rutgers University

Harper, G., Brownlee, A., Hall, T. E., Seymour, R., Lyons, R., and Ledwith, P. 2003. *Global Progress toward Transgenic Food Animals: A Survey of Publicly Available Information*. Wellington, New Zealand: Food Standards Australia New Zealand. Retrieved from //www.foodstandards.gov.au/_srcfiles/Transgenic%20Livestock%20Review%20CSIRO%20FINAL%2012Dec20031.pdf. Accessed May 3, 2007.

Hoban, T. 2002. Education Required for Animal Biotechnology. Paper presented for the College of Agriculture and Life Sciences, North Caroline State University. Retrieved from www4.ncsu.edu/~hobantj/biotechnology/biotechnology_webpage.html. Accessed June 17, 2005.

Hoban, T. 2003. Public Perceptions of Biotechnology. Retrieved from www4.ncsu.edu/~hobantj/biotechnology/biotechnology_webpage.html. Accessed June 17, 2005.

Hoban, T., and Kendall, P. 1993. *Consumer Attitudes about Food Biotechnology*. Raleigh, N.C.: North Carolina Cooperative Extension Service. Retrieved from www4.ncsu.edu/~hobantj/biotechnology/biotechnology_webpage.html. Accessed June 17, 2005.

Hoban, T., Woodrum, E., and Czaja, R. 1992. Public Opposition to Genetic Engineering. *Rural Sociology* 57(4): 476–495.

Hossain, F., and Onyango, B. 2004. Product Attributes and Consumer Acceptance of Nutritionally Enhanced Genetically Modified Foods. *International Journal of Consumer Studies* 28(3): 255–267.

Huffman, W., Shogren, J., Rousu, M., and Tegene, A. 2003. Consumer Willingness to Pay for Genetically Modified Food Labels in a Market with Diverse Information: Evidence from Experimental Auctions. *Journal of Agricultural and Resource Economics* 28: 481–502.

Jongbloed, A. W., and Lenis, N. P. 1998. Environmental Concerns about Animal Manure. *Journal of Animal Science* 76(10): 2641–2648.

Kelso, D. D. T. 2004. Genetically Engineered Salmon, Ecological Risk, and Environmental Policy. *Bulletin of Marine Science* 74(3): 509–528.

Martin, A. November 12, 2003. One Fish, Two Fish, Genetically New Fish. *Chicago Tribune*, 1.

Mench, J. A. 1999. Ethics, Animal Welfare, and Transgenic Farm Animals. In *Transgenic Animals in Agriculture*, ed. J. D. Murray, G. B. Anderson, A. M. Oberbauer, and M. M. McGloughlin. Oxford: CABI Publishing, 251–268.

Minnesota Daily. October 30, 2001. Enviropigs Will Not Help Environment [Editorial].

Muir, W., and Howard, R. 1999. Possible Ecological Risks of Transgenic of Transgenic Organism Release When Transgenes Affect Mating Success: Sexual Selection and the Trojan Gene Hypothesis. *Proceedings of the National Academy of Sciences of the U.S.A.* 96(24): 13853–13856.

National Research Council. 2004. *Animal Biotechnology: Science-Based Concerns*. Washington, D.C.: National Academies Press.

Pew Initiative on Food and Biotechnology. September 2003a. An Update on Public Sentiment about Agricultural Biotechnology. Mellman Group, Inc./Public Opinion Strategies for the Pew Initiative. Retrieved from pewagbiotech.org/research/2003update/2003summary.pdf. Accessed May 1, 2007.

Pew Initiative on Food and Biotechnology. September 2003b. GM Food Safety: Are Government Regulation Adequate? Retrieved from pewagbiotech.org/buzz/display.php3?StoryID=42. Accessed July 13, 2004.

Pew Initiative on Food and Biotechnology. 2004a. Issues in the Regulation of Genetically Engineered Plants and Animals. Retrieved from pewagbiotech.org/research/regulation/Regulation.pdf. Accessed July 13, 2004.

Pew Initiative on Food and Biotechnology. August 2004b. Genetically Modified Crops in the United States. Retrieved from pewagbiotech.org/resources/factsheets/display.php3?FactsheetID=2. Accessed July 14, 2005.

Post, R. July 25, 2005. E-mail re: Information on U.S. Regulations Regarding Labeling of Animal Food Products. Director, Labeling and Consumer Protection Staff, Food Safety and Inspection Service, USDA.

Poultry Products Inspection Act of 1957. U.S. Public Law 172. 85th Cong., 1st Sess., 28 August 1957.

Pursel, V. G., Pinkert, C. A., Miller, K. F., Bolt, D. J., Campbell, R. G., Palmiter, R. D., Brinster, R. L., and Hammer, R. E. 1989. Genetic Engineering of Livestock. *Science* 244(4910): 1281–1288.

Reiss, M., and Straughan, R. 1996. *Improving Nature? The Science and Ethics of Genetic Engineering*. Cambridge: Cambridge University Press.

Rollin, B. 1995. *The Frankenstein Syndrome: Ethical and Social Issues in the Genetic Engineering of Animals*. Cambridge: Cambridge University Press.

Rubel, A., and Streiffer, R. 2005. Respecting the Autonomy of European and American Consumers: Defending Positive Labels on GM Foods. *Journal of Agricultural and Environmental Ethics* 18(1): 75–84.

Russo, J. E., Metcalf, B. L., and Stephens, D. 1981. Identifying Misleading Advertising. *Journal of Consumer Research* 8(2): 119–131.

Sigurdson, C. 1995. Purdue's "Kindler, Gentler Chicken" Moves into Real-world Test. *Feedstuffs* (January 16): 47–48.

Smil, V. 2000. Phosphorus in the Environment: Natural Flows and Human Interferences. *Annual Review of Energy and the Environment* 25(1): 53–88.

Streiffer, R., and Hedemann, T. 2005. The Political Import of Intrinsic Objections to Genetically Engineered Food. *Journal of Agricultural and Environmental Ethics* 18(2): 191–210.

Streiffer, R., and Rubel, A. 2004. Democratic Principles and Mandatory Labeling of Genetically Engineered Food. *Public Affairs Quarterly* 18(3): 223–248.

Union of Concerned Scientists. 2001. Backgrounder: Genetically Engineered Salmon. Retrieved from www.ucsusa.org/food_and_environment/biotechnology/page.cfm?pageID=327. Accessed on June 17, 2005.

U.S. Congress, Office of Technology Assessment. 1987. *New Developments in Biotechnology—Background Paper: Public Perceptions of Biotechnology*. Paper OTA-BP-BA-45. Washington, D.C: U.S. Government Printing Office.

Varner, G. 2002. Animals. In *Life Science Ethics*, ed. Gary Comstock. Ames: Iowa State Press, 141–168.

Vestel, L. B. October 26, 2001. The Next Pig Thing. *Mother Jones*.

# 6

# Mandatory Genetic Engineering Labels and Consumer Autonomy

*Peter Markie*

Proponents of mandatory labels for GE (genetically engineered) products (Jackson 2000; Streiffer and Rubel 2004; Rubel and Streiffer 2005) often present what I call the autonomy argument: The labeling of GE products as such should be mandatory, since consumers need this information to make purchase decisions that are informed by their values. The argument is attractive, for it promises to settle the issue of mandatory labels for GE products without requiring us to first resolve difficult issues of their safety.[1] The argument is also an obvious non sequitur. We thus face two questions: (1) How can we best develop the argument? (2) Is the argument, suitably developed, successful? In this chapter I examine three attempts to develop the argument, one of my own design and two that have been offered by others. I show that each is unsuccessful. I conclude with some general comments about the poor prospects for a successful version of the argument.

## CONSUMER AUTONOMY AND THE OBLIGATIONS OF GOVERNMENTS AND MARKETERS

The first version of the autonomy argument appeals to consumer autonomy to support the marketers' responsibility thesis:

> Marketers' responsibility thesis: The marketers of GE products have a moral obligation to label GE products as such for consumers.

It then appeals to the importance of this obligation and the slim chance of its being honored without legal regulation to derive the mandatory labeling thesis:

> Mandatory labeling thesis: Governments of societies in which GE products are marketed are morally obligated to legally require that those marketing these products label GE products as such for consumers.

Neither thesis is satisfied by negative ("contains no GE ingredients") labels. Each is neutral on the precise content of the required, positive labels, as well as on the level of GE ingredients sufficient to require a mandatory label.

Rubel and Streiffer (2005: 80) present what seems to be a statement of the first stage of the argument, the inference from consumer autonomy to the marketers' responsibility thesis:

> When people desire information because of deeply held values, the obligation to promote autonomy (and therefore incur expense) is substantial. This implies an obligation to incur expense promoting autonomy in the case of GM [genetically modified] labels, for the reasons consumers desire GM labels are often at the core of their values. Many people are concerned about whether the food they eat contains genes from foods that are prohibited in their religion or culture; vegetarians may be concerned that non-animal food products contain animal-derived genes; animal welfare supporters may be concerned about the impact of genetic engineering on animal well-being; "latent purificationists" ... may think that genetic modifications make food less safe, good, or pure; and some people wish to avoid GM foods to support small or organic farms, even if they have no concerns about consuming GM foods per se. ... These reasons are related to persons' deeply held values, and therefore providing the information necessary for them to act on those values is essential to respecting consumer autonomy in a way that goes beyond what is indicated by the raw number of consumers who care.

While Rubel and Streiffer do not say just who has the moral obligation to promote consumer autonomy in the GE market, it is reasonable to believe they have the marketers of GE products in mind.

Kristen Hansen (2004: 68–69) outlines the next stage of the argument, which takes us from the moral obligations of GE marketers (the marketers' responsibility thesis) to the moral obligations of legal regulators (the mandatory labeling thesis):

> The purpose of mandatory labeling must be to protect the consumers' ability to choose knowledgeably and thereby respect the consumers' autonomy. Such protection, the argument seems to go, should not be left to the free market. This is because the free market cannot be counted on to serve such a purpose, except in cases where there is a majority consumer demand for the products in question. If only a few consumers care about keeping their diets GM free, producers will not voluntarily give them the information by labeling their products. Therefore mandatory labeling is required in order to assure consumers' autonomy.[2]

Neither market forces nor, presumably, their own consciences will guarantee that those marketing GE products honor their ethical duty to label their goods as such. Legal regulation is therefore required to ensure that they do so.

Each stage of the autonomy argument needs to be developed further. The remarks by Rubel and Streiffer, on the one hand, and Hansen, on the other, at best sketch the reasoning of each stage. I focus just on the first stage, the argument from considerations of consumer autonomy to the marketers' responsibility thesis. I present what I take to be the most promising way to develop it and show where it goes wrong. This first version of the autonomy argument breaks down at the start.[3]

## FROM THE AUTONOMY OF CONSUMERS TO THE MORAL OBLIGATIONS OF MARKETERS

The argument for the marketers' responsibility thesis employs some reasonable assumptions. First, relevant information for a consumer is information that provides the consumer with a reason to believe that a particular product choice does or does not promote his or her values. Second, autonomy comes in degrees, and all other things being equal, consumers are more, rather than less, autonomous when their purchase decisions are informed by information that is relevant for them. Third, autonomy is a good. If our following a particular course of action will increase the autonomy of others, we have a prima facie duty, one of beneficence, to follow that course of action. Fourth, that our providing others with some information will increase their autonomy is not sufficient to give them a moral right against us to our providing them with that information. This last point merits emphasis because it is tempting to assume that consumers have a right to whatever information will promote their autonomy. Not so. Car dealers will promote their customers' autonomy by telling them when the same automobile can be purchased for less elsewhere or as part of an upcoming sale, assuming that the customers value paying the lowest price possible. Customers clearly do not have a right against dealers to be provided with this information, however.

Relative to these assumptions, here is the first part of a two-part argument from consumer autonomy to the marketers' responsibility thesis:

Autonomy Argument for Marketers' Responsibility Thesis (Stage 1)

1. If, all other things being equal, gaining some information will increase the autonomy of some consumers' purchase decisions and those marketing the product are able to provide them with that information by adopting practice X, then it is prima facie morally obligatory that those marketing the product adopt X.
2. All other things being equal, some potential consumers of GE products will increase the autonomy of their purchase decisions by gaining the information that the GE products are GE, and those marketing the GE products are able to provide them with that information by labeling GE products as such.
3. Therefore, it is prima facie obligatory that those marketing GE products provide potential consumers of their products with the information that the products are GE by labeling them as such.

The first premise follows from the initial assumptions. The second premise asserts that there is at least one potential consumer of GE products whose values make the information that a product is GE relevant. This fits with the view of at least some proponents of the autonomy argument that "[c]onsumer autonomy is about individuals, and about whether individuals have the information they think is necessary to make important decisions" and that "the number that care is irrelevant to whether or not labels protect their autonomy" (Rubel and Streiffer 2005: 76–77).[4] I shall assume that the second premise is true for all the societies of interest to us. This initial stage of the argument from consumer autonomy to the marketers' responsibility thesis is sound.

The argument's conclusion is quite minimal, however. Prima facie duties come cheap. Our all-things-considered duties are the ultimate moral standard for our behavior. How are we to get from a prima facie duty for marketers to label their GE products to an all-things-considered duty for them to do so? To my knowledge, proponents of the autonomy argument do not address this question. We have to fill the gap for them:

**Autonomy Argument for Marketers' Responsibility Thesis (Stage 2)**

1. It is prima facie obligatory that those marketing GE products provide potential consumers of their products with the information that the products are GE by labeling them as such.
2. If it is prima facie obligatory that those marketing GE products do X and there is no morally sufficient reason for them not to do so, then they are morally obligated to do so.
3. There is no morally sufficient reason for those marketing GE products not to provide potential consumers with the information that their products are GE by labeling them as such.
4. Therefore, those marketing GE products are morally obligated to provide potential consumers of their products with the information that the products are GE by labeling them as such (the marketers' responsibility thesis).

The first premise is the conclusion of the first stage of the argument. The second and third premises contain the concept of a morally sufficient reason not to act on a prima facie duty. I shall not try to define this concept, but I can clarify it some. Consider four different ways in which we might gain a morally sufficient reason not to honor a prima facie duty to share some information.

One morally sufficient reason not to honor a prima facie duty is the existence of a competing, equally or more stringent prima facie duty. If I have a prima facie duty to promote your autonomy by providing you with some information but my promise to keep the information confidential gives me an equal or more stringent prima facie duty, then I have no all-things-considered duty to provide you with the information.

A second morally sufficient reason not to honor a prima facie duty to share information in a certain way is our ability to equally well respect the source of the duty through an alternative course of action. A car dealer will increase the autonomy of his customers by informing them of the repair history of his used cars. The practice of posting each car's repair history on its side window is one way to provide the information, so he has a prima facie duty to do so. Yet, he can also respect the source of his duty by simply posting a sign on each car's window that says "Repair history available on request" and making the information readily available to any customer who asks. The existence of this alternative keeps the car dealer from having an all-things-considered obligation to take the first option. Either option is morally permissible for him.

A third morally sufficient reason to forgo a prima facie duty to share information in a certain way is the implementation by others of a practice that equally well satisfies the basis for our prima facie duty. Marketers of nonkosher products have a prima facie obligation to promote the autonomy of their customers by labeling their products as nonkosher. They have no all-things-considered obligation to do so, however. The existing system of labeling kosher products as kosher promotes consumers' autonomy sufficiently well. Consumers are able to make choices informed by their values, by simply assuming, as is likely, that whatever is not labeled kosher is not kosher. The absence of a label of kosher effectively enough serves the same purpose as the presence of a label of kosher.

A fourth morally sufficient reason not to honor a prima facie duty to share information stems from the subject's rights. Consider a company that sells a sauce made with a "secret formula." The autonomy of some consumers might be increased by knowledge of the formula; perhaps they value purchasing (or avoiding) products that have the precise distribution of ingredients detailed in the formula or perhaps they value purchasing products where they have "insider knowledge" of any secret formulas. The company nonetheless has a morally sufficient reason not to provide the information. Its right to the formula entitles it to withhold the information in the absence of any more stringent consumer right to it. It does nothing wrong if it refuses to surrender one of its resources, in this case its monopoly on knowledge of the formula, to promote consumer autonomy.

There are probably other morally sufficient reasons not to honor a prima facie duty to promote autonomy by sharing information, but a consideration of these four enables us to understand the argument's second and third premises sufficiently well for our purposes. I shall grant, for the sake of argument, the second premise that prima facie duties become all-thing-considered duties in the absence of a morally sufficient reason to do otherwise. I want to focus on the third premise and the issue of whether the marketers of GE products have a morally sufficient reason not to promote consumer autonomy by labeling their products as such. I consider four attempts to show that they have a sufficient reason. The first is unsuccessful. The second and third show important ways in which this version of the autonomy argument needs to be developed still further. The fourth shows that the argument is simply unsound. At least as I construct it here, the inference from consumer autonomy to the marketers' responsibility thesis fails.[5]

## FROM THE AUTONOMY OF CONSUMERS TO THE MORAL OBLIGATIONS OF MARKETERS: OBJECTIONS

One objection appeals to the third reason I cite above for not honoring a prima facie duty: The implementation by others of an alternative practice equally well serves the end of promoting consumer autonomy. The marketers of nonkosher foods have no all-things-considered duty to label their products; the production and labeling of kosher alternatives sufficiently well promote consumer autonomy. Analogously, opponents of GE labels may claim, consumer autonomy is equally well promoted by the development and labeling of non-GE alternatives, thus relieving the marketers of GE products of an all-things-considered duty to label their products.

This reply misses a fundamental point. The issue is whether, in the current context, the marketers of GE products have a morally sufficient reason not to honor their prima facie duty to label. The actual existence of a readily available, broad supply of labeled non-GE alternatives might be such a reason. Even the likely development of such a supply of labeled non-GE alternatives might be sufficient. Yet, no such supply of labeled non-GE alternatives exists or is likely to exist in the near future. The closest we have is a quite limited supply of labeled organic products, which by implication are non-GE. The mere fact that an adequate supply of labeled non-GE alternatives might exist sometime in the future does not excuse marketers today from honoring their prima facie obligation to promote the autonomy of the consumers to whom they are marketing GE products now.

A second objection concerns whether the marketers of GE products have a more stringent prima facie duty that balances out or overrides their prima facie duty to label their products. Whether GE labels will promote the autonomy of any particular consumer depends in part on the consumer's ability to combine the information on the label with an adequate understanding of the issues surrounding GE products to make purchase decisions that are informed by his or her values. If some consumers have a poor or confused understanding of the issues surrounding GE products, positive GE labels may actually lead them to make decisions that do not best promote their values. The very GE labels that may increase the autonomy of some well-informed consumers may also lessen that of poorly informed ones. The prima facie obligations of the marketers of GE products to label them in order to promote the autonomy of some consumers may be balanced by an equally, or even more, stringent prima facie obligation not to label them in order not to lessen the autonomy of others. Proponents of this version of the autonomy argument need to develop the argument to meet this concern. They must show that positive GE labels, in the context of consumers' current understanding of GE products, will significantly promote, rather than lessen, overall consumer autonomy, or they must change their claim to the assertion that the marketers of GE products are obligated not only to label their products but also to effectively educate consumers on their meaning.

A third objection appeals to the second morally sufficient reason for not honoring a prima facie duty: the agent's ability to honor the source of the duty by an alternate means. While the marketers of GE products can promote consumer autonomy by labeling their products as such, they can also accomplish the same end though other means, say, by developing websites on which they identify their GE products and the GE nature of each. This way of promoting consumer autonomy requires a bit more of consumers, as they have to seek out the information they desire, but it also has an increased benefit for consumers, as marketers can provide more information than is available through a product label.[6]

The success of this objection depends on whether there is a morally significant difference between product labels and other ways of providing the same information to consumers. The burden of proof is on those offering this version of the autonomy argument. To establish their premise that marketers of GE products have no morally sufficient reason not to label their products, they must show that a morally significant difference favors labels to the point where the marketers' use of any other means of providing consumers with the information would be morally wrong. It is not enough for them to show that consumers prefer labels, or that labels are more "consumer friendly," or that labels are a more traditional way of providing consumer information. To show that the marketers of GE products are obligated to use GE labels, they must produce considerations that show that it would be morally wrong for marketers to employ any other means of communication instead. How supporters of the autonomy argument might show this is unclear, but the success of their argument hinges on their doing so.

A final and fatal objection to this version of the autonomy argument focuses on the fourth morally sufficient reason for not honoring a prima facie obligation. Our rights can entitle us to set aside a particular prima facie duty of beneficence. Faced with multiple opportunities to give to charity, we have a prima facie duty of beneficence to do each, but our right to our property entitles us to do some but not others. Analogously, marketers of GE products have multiple opportunities to promote the autonomy and, more generally, the welfare of consumers. They can do so by providing a variety of worthwhile products, they can do so by contributing to various social causes and initiatives, and they can do so by labeling GE products as such. They have a prima facie duty of beneficence in each case. As long as consumers do not have a right to their pursuing a particular one, the marketers' right to their resources entitles them to select some ways to advance the consumers' welfare and not others. In particular, they are entitled not to promote consumer autonomy by labeling their GE products as such.

Consider an analogous case. Since some consumers desire to purchase only "dolphin-safe" tuna, marketers of tuna that is not "dolphin safe" have a prima facie duty to promote consumers' autonomy by labeling their product as such. Yet, they have no all-things-considered obligation to do so. They do not act unethically, if they market their tuna without any such label. It is not that the availability of labeled, "dolphin-safe" tuna equally well promotes consumer autonomy. Even before the existence of labeled, "dolphin-safe" tuna, the marketers of tuna that was not "dolphin safe" were not ethically obligated to label their product as such. They lack an all-things-considered obligation because instituting a labeling program involves additional expense and risk, and they are morally entitled to decide whether they wish to invest their resources to honor that particular prima facie duty of beneficence.

Other examples abound. Given current consumer values, marketers have the opportunity, and so the prima facie duty, to promote consumer autonomy by sharing all sorts of product information through labels, for example, whether or not the product contains any components produced in the United States (or the European Union, Israel, or any of a number of countries of interest to some consumers), whether or not the product is produced by union labor, and whether or not it was produced in an environmentally friendly way, as such production is defined by any particular group of interested consumers. While marketers have a prima facie duty to promote consumers' autonomy by sharing such information, they certainly do not have an all-things-considered duty to do so. The reason again is that, in the absence of a consumer right to the information, marketers are morally entitled to determine whether they want to take on the expense and risk associated with providing it.

Might consumers have a right to know that GE products are genetically engineered? Consumers could gain such a right though establishing an agreement with marketers that they, the marketers, will provide the information. No such agreement currently exists, however. Consumers might derive the right from another one, but it is hard to find a plausible candidate for such a more basic right. It will not do to appeal to consumers' right not to be deliberately deceived. Marketers who do not label their GE products as such are not deliberately deceiving consumers. They do not represent their GE products as being non-GE, any more than the marketers who fail to label products made in China represent them as made elsewhere.

Yet, what if most consumers believe, falsely, that they are buying non-GE products and rely on this information in making their purchase decisions, and GE marketers know all the while that they are making their purchase decision on this false belief? Don't the marketers then violate the consumers' right not to be deceived, if they fail to inform consumers that the products are actually GE?[7] The answer, again, is no, as long as the marketers are not responsible for the creation of consumers' false belief in the first place. Suppose we purchase a car relying, in part, on our false belief that it will not be on sale in two weeks. The car dealer knows of the upcoming sale, knows of our ignorance, and knows that our ignorance plays a role in our purchase decision. As long as he is not responsible for the existence of our false belief, for example, he has not told us that there will be no sale in the near future or posted signs advertising his current prices as the lowest for the next six months, he does not unethically deceive us by not informing us of the upcoming sale. The same is true, by analogy, of the marketers of GE products. As long as they are not responsible for consumers' false belief that products are non-GE, they do not unethically deceive consumers by failing to correct it. If anything, the case for charging the marketer of GE products with unethical deception is even weaker than that for charging the car dealer, for the consumers of GE products bear greater responsibility than the typical car buyer for their ignorance. Even proponents of the autonomy argument admit that information on the GE nature of most products is readily available to consumers willing to seek it out, as already noted. Car buyers do not have such easy access to the pricing plans of particular car dealers.[8]

In all, this first version of the autonomy argument is unsuccessful. The appeal to consumers' autonomy supports a prima facie duty for the marketers of GE products to label their products, but it doesn't support an all-things-considered obligation for them to do so. I now want to consider another version of the argument, one that is focused on the obligations of the FDA, in particular.

## CONSUMER AUTONOMY AND THE FDA

Streiffer and Rubel (2004: 226) state the argument as follows:

> The decision as to whether GE foods should be labeled is within the discretion of the FDA. In order to decide how to exercise its discretion on this matter, the FDA should ask what the best justification is for the existing labeling provisions, and then determine whether or not that justification also supports a labeling requirement for GE foods. The best justification for the existing labeling statutes gives a primary

> place to consumer autonomy, and considerations of consumer autonomy support a labeling requirement for GE foods. Thus, the FDA should require labels on GE foods.

Staying close to the text of their statement, we have the following first approximation of their reasoning:

**The Consumer Autonomy/FDA Argument (Version 1)**

1. The FDA has the discretion to require the labeling of GE foods.
2. In exercising its discretion in this matter, the FDA should apply the principles contained in the best justification for its existing labeling provisions.
3. The principles in the best justification for its existing labeling provisions give a primary place to consumer autonomy.
4. Considerations of consumer autonomy support a labeling requirement for GE foods.
5. Therefore, the FDA should require labels on GE foods.

Several aspects of the argument need to be clarified before it can be properly evaluated. The logical form of the argument is unclear—just how does the conclusion follow from the premises? Premise 3 refers to the principles in the best justification for current FDA labeling requirements—what are those principles? Is the argument concerned with all of the FDA's existing requirements or only with those it has a moral obligation to adopt? Presumably, we do not want to appeal to any labeling provisions that are morally mistaken to support the claim that a new labeling provision is morally obligatory. Premise 3 refers to a "primary place" for consumer autonomy in the best justification of existing FDA labeling requirements—just what is a primary place? If the argument is going to show that consumer autonomy itself is sufficient to justify the mandatory labeling of GE products, the primary place of consumer autonomy in the justification of existing labeling requirements is presumably that of a sufficient condition for mandatory labels.

I suggest that the best way to address these points is to restate Rubel and Streiffer's argument as follows:

**The Consumer Autonomy/FDA Argument (Version 2)**

1. The FDA has the legal and moral authority to require labels for GE foods.
2. The FDA will promote consumer autonomy by requiring labels for GE foods.
3. If the FDA has the legal and moral authority to require labels for a product and it will promote consumer autonomy by doing so, then the FDA is morally obligated to do so (the FDA/consumer autonomy principle).
4. Therefore, the FDA is morally obligated to require labels for GE foods.

The first premise is based on the assumption that the enabling legislation defining the FDA's authority is morally and legally justified and extends the FDA's powers to the case of GE foods. Rubel and Streiffer defend the second premise by providing empirical information regarding the extent of consumer interest in knowing the GE nature of food products. The justification for the third premise, which I call the FDA/consumer autonomy principle, is that it is the best explanation of the FDA's assumed moral obligation to impose other labeling requirements: The FDA has a moral obligation to impose labeling requirements in those other cases simply because the requirements promote consumer autonomy.

I shall grant the first two premises and focus on the FDA/consumer autonomy principle.[9] An important aspect of the principle is unclear. It offers the autonomy of consumers as a sufficient condition for mandatory FDA labels with no specification of what percentage or number of consumers must have their autonomy advanced by labels and no account of how strongly they must value the information to be provided by labels. Surely the fact that a mandatory label would advance the autonomy of one consumer who barely values having the information is not sufficient to justify an FDA labeling requirement with its accompanying economic costs and its limitations on liberty. Rubel and Streiffer do not address these points, however. To fill the gap, I shall assume that we are interpreting the principle in terms of some significant number/percentage of consumers and some significant level of value.

Even with this background assumption in place, we have good reason to be suspicious of the principle. It implies, quite implausibly, that the FDA is morally required to mandate labels whenever its doing so will promote consumers' autonomy, no matter how irrational the beliefs that underlie their values and no matter how unrelated or even opposed to society's overall interests those values, their promotion, or the labeling requirement itself may be. In offering consumer autonomy as a sufficient condition for mandatory labeling, it claims that consumer autonomy outweighs any of the other considerations we regularly take to be relevant to a labeling decision, no matter how strong those other considerations may be. This may be correct, but it certainly seems quite unlikely at the start. Consumer autonomy is one good, but it is not the only one. The principle needs strong support.

The support for the FDA/consumer autonomy principle is an inference to the best explanation. We are to accept it and apply it to the case of GE products because it provides the best explanation of other cases where the FDA clearly has a moral obligation to require labels. Before I discuss the details of this argument, it is worth pausing to consider just what a successful inference to the best explanation requires in this case. Since the FDA/consumer autonomy principle presents consumer autonomy as sufficient in itself to obligate the FDA to impose mandatory labels, the argument must present cases that have two characteristics. First, the FDA must have an all-things-considered obligation to impose mandatory labels. Second, an appeal to consumer autonomy alone must be not only an explanation but also the best of any competing explanations for the FDA's obligation. With this in mind, consider the cases presented by Rubel and Streiffer. In each case, the FDA is clearly obligated to impose a labeling requirement, and considerations of consumer autonomy plausibly play some role in explaining why. The problem is that other considerations also play important, and apparently necessary, roles in the FDA's obligation.

The first labeling provision that Streiffer and Rubel (2004: 229) consider is the requirement that food labels carry the name and place of business of the food's manufacturer, packager, or distributor:

> What is the best justification of this requirement? Although it is true that this information could facilitate safety in cases of suspected food-borne illness, this can't be the only consideration motivating it, for there are more efficient ways to track suspect foods than by requiring that information (e.g., requiring a phone number and lot number would suffice). Rather the best justification is that place of manufacture matters to people in making purchasing decisions; their values either support making food choices based upon the information, or they wish to make their decisions with that information readily available, regardless of whether the information directly influences any particular choice.

Nothing in this description of the case shows that the promotion of consumer autonomy is the best explanation of the FDA's obligation to require manufacturer identification labels. For all Rubel and Streiffer tell us, the labeling requirement may be best explained by a combination of considerations ranging from consumer autonomy to safety concerns, which they admit to be present, to the value of promoting the accountability of manufacturers for the quality of their products. The best explanation of the FDA's obligation may also involve considerations about the cost of the labeling provision for both manufacturers and consumers.

Streiffer and Rubel (2004: 229–230) next consider quantity labels:

> The FDCA also requires a statement of food quantity "in terms of weight, measure, or numerical count." One might argue that the FDCA requires an accurate statement of quantity only to ensure that customers are not fraudulently misled by, for example, large cereal boxes or heavy packaging. This is of course true but only reinforces the point about autonomy. It is necessary that an autonomous choice be made free of deception; thus rules requiring full disclosure of quantity and thereby preventing (or mitigating) deceptive packaging serve precisely to promote autonomy. One might also argue that the justification for such a provision is that it protects the economic interests of consumers. Presumably it does, but nonetheless it is still best explained as a mechanism for protecting consumer autonomy. More importantly, requiring information about quantity would be justified even if producers did not induce consumers to make economically detrimental decisions by deceptive packaging. Thus, the best justification for requiring accurate quantity labels is that it respects consumer autonomy.

Even as Rubel and Streiffer admit that the basis for requiring the labels in the first instance is to prevent fraud and promote the economic interests of consumers, they claim, that these goals are themselves justified by the more fundamental end of providing consumers with whatever information they value in making purchase decisions. Moreover, even if all producers were honest and did not engage in deceptive packaging, the FDA would still be obligated to require quantity labels, and this fact, they claim, can be explained only by the principle that the promotion of consumer autonomy is sufficient to obligate the FDA to require labels.

Yet, there is at least one other equally plausible explanation of the FDA's obligation. It is required to mandate quantity labels in order to ensure that the relation between sellers and purchasers is fair. Sellers offer their product in exchange for a precise amount of money. Fairness requires that purchasers know how much product they are getting for their money, just as sellers know how much money they are getting for their product. The FDA is obligated to promote fairness for its own sake. Moreover, in adopting a labeling requirement in support of fairness, the FDA must balance the benefits to be gained against the costs of implementing the requirement for both marketers and consumers, and in this case, the gains in terms of fairness outweigh the costs of implementation. This explanation supports mandatory labels even in the absence of deceptive packaging. Rubel and Streiffer need to show that this alternative explanation of the FDA's obligation is somehow inferior to their appeal to consumer autonomy as sufficient by itself to support the FDA's obligation.

Streiffer and Rubel (2004: 230) next consider the requirement that labels be truthful and nonmisleading:

> Consider the FDCA's general requirement that labels be truthful and non-misleading. As noted above, autonomy requires that one be able to incorporate one's values into choices as one sees fit. Where a label is deceptive by being untruthful or misleading, there is a sense in which the choice is not one's own at all. Deception allows another person to manipulate one's decisions and thereby fails to respect autonomy. Requirements for truthfulness and non-misleadingness therefore serve to respect autonomy.

Rubel and Streiffer are clearly right that deceptive labels limit consumer autonomy, but that doesn't mean that the FDA/consumer autonomy principle best explains why the FDA is obligated to ban them. An alternative principle offers an equally plausible account: When not outweighed by the costs for various stakeholders, the fact that a labeling requirement is necessary to prevent marketers from deceiving consumers is sufficient to obligate the FDA to impose it; deception is to be prevented for its own sake. This principle explains the FDA's obligation in the case at hand, but it does not support an FDA obligation to mandate labels for GE products. Such labels are not necessary to prevent marketers from deceiving consumers, for reasons already considered.

Streiffer and Rubel (2004: 230) offer safety and nutrition labels as their last example:

> It might strike some as odd to say that considerations of autonomy, as opposed to considerations of health, justify labeling on the basis of safety and nutrition, but what explains the importance of health? It makes sense to accommodate health-based food decisions not because health concerns are intrinsically more important than other sorts of concerns. Rather, the desire for health-related information is widely shared, both because people value health itself and because health is instrumentally necessary to whatever sort of life one wishes to lead. Thus, even safety and nutrition labeling is ultimately justified because it promotes autonomy by providing consumers information important to them.

The argument here is an obvious non sequitur. That the FDA should require safety and nutrition labels is clear. That consumers widely desire such information is also clear. That the latter best explains the former, as Rubel and Streiffer conclude, simply does not follow. There are other plausible explanations. An obvious one is that society has a strong interest, if only in terms of managing health care costs, in promoting health-based purchase decisions, and the FDA should use its discretion to further this interest when the costs associated with a particular way of doing so do not outweigh the benefits.[10] Rubel and Streiffer give us no reason to believe that society has a similar interest in promoting GE-based purchase decisions.

In all four cases, Rubel and Streiffer fail to show that the FDA's moral obligation derives from the principle that, given the legal and moral authority to adopt a particular labeling provision, the FDA should do so whenever the provision will promote consumer autonomy. There is an important lesson to draw from our consideration of these cases. Almost any labeling requirement can be linked to the end of promoting consumer autonomy by the following reasoning: This labeling requirement is intended to provide consumers with information; the information is valuable to them relative to promoting their autonomy; hence, the best explanation for requirement must be that consumer autonomy itself is sufficient to support an obligation to mandate labels. The problem is that, while the premises are true, the conclusion does not follow.

## CONSUMER AUTONOMY AND DEMOCRATIC VALUES

A third way of arguing for mandatory GE labels on the basis of the values and preferences of consumers is the democratic equality argument. Rubel and Streiffer (2004: 234) state it as follows:

> The decision as to whether GE foods should be labeled is a collective decision within the purview of the citizenry. And although Congress is frequently justified in delegating a decision within the purview of the citizenry to a scientific agency with the authority to overrule citizens' preferences, this is not the case with labeling. So, if the public wants labeling (and if the FDA does not provide it), then Congress should require it. Further, when consumers are informed about the prevalence of GE food products and ingredients, a vast majority say that there should be labeling. It follows that Congress should require labeling.

It is again worth considering their reasoning in detail:

**The Democratic Equality Argument**

1. If the following conditions are met with regard to a proposed labeling requirement, then Congress is obligated to impose the labeling requirement:
   - i. The public wants the labels to be required,
   - ii. The decision to require the labels is a collective decision within the purview of the citizenry, and
   - iii. Congress is not justified in delegating the labeling decision to an agency with authority to overrule citizens' preferences.
2. The public wants GE products to be labeled as such.
3. The GE labeling decision is a collective decision within the purview of the citizenry.
4. Congress is not justified in delegating the GE labeling decision to an agency with authority to overrule citizens' preferences.
5. Therefore, Congress is obligated to impose the GE labeling requirement.

I call the first premise of the argument the democratic equality principle. Rubel and Streiffer do not argue for it; they instead focus their attention on each of the subsequent premises. They cite various studies of consumer preferences in support of premise 2. They argue for premise 3 on the ground that "following the majority's will to require labeling would neither impinge upon people's fundamental rights nor undermine institutions and procedures necessary for democracy" (Streiffer and Rubel 2004: 236); they take the U.S. Constitution as the relevant standard for determining whether fundamental rights are violated. They offer several considerations in support of premise 4: The public has not consented to such a congressional delegation of authority, such a delegation is not justified on paternalistic grounds, and if Congress follows public opinion in this case, rather than overruling it on its own or through a delegation of an authorized agency, it will not treat any stakeholders unfairly.

Before evaluating the argument, it is worth noting just how far I have come in my attempt to show that consumer autonomy justifies mandatory GE labels. The initial attempt was centered on the idea that considerations of consumer autonomy give the manufacturers of GE products a moral obligation to label their products. Even if consumers do not have a right to the information, marketers have a duty of beneficence to provide it. If marketers were to ask why they are being required to label their products, the answer would be that they have an important moral obligation of beneficence to do so. The current argument makes no such pretence of basing a GE labeling requirement in the good of consumers and the moral obligations of the marketers of GE products. It attempts to base the requirement simply on consumers' desires. If manufacturers were to ask why they are being required to label their products, the answer would simply be that consumers want them to do so. Having set aside the appeal to the consumers' "right to know" and even the appeal to the consumers' welfare, we have come to the consumers' right to impose their will on the marketers of GE products under democratic principles.

A major defect in the argument becomes clear once we take a close look at the democratic equality principle. Whether or not Congress has a moral obligation to impose a labeling requirement in response to the public's desires depends in part on whether the requirement is fair to those who would be forced to assume its burdens. As Rubel and Streiffer explain the democratic equality principle, it is sensitive to considerations of fairness in two ways. First, the principle's second condition will not be met, and premise 3 in the argument will not be true, if the labeling requirement violates fundamental constitutional rights. This provision captures some concerns regarding fairness but not all. (Not all demands of fairness are constitutional requirements.) It certainly does not capture those involved in the present case. Second, the principle's third condition will not be met, and premise 4 in the argument will not be true, if the labeling provision treats the interests of some stakeholders unfairly in a way serious enough to overrule the public's desire for labels. This is clearly relevant to the present case. A major objection to mandatory GE labels is just that they will unfairly impose a costly burden on some stakeholders in order to provide some consumers with information in the absence of any showing that those consumers have a preexisting right to the information or that making its display mandatory is the best way to promote an important social good. Rubel and Streiffer acknowledge this objection in defending their fourth premise. The problem is that they fail to show that it is mistaken. They tell us the following:

> Labeling would, in all likelihood, deprive some producers of GE foods of some of their profits. Thompson takes this as grounds for objecting to mandatory labels, saying that they would "groundlessly reduce the commercial viability of genetically engineered foods and this could plausibly be interpreted as interference in the rights of the food industry." However, producers are not entitled to profits resulting from consumer ignorance of facts about their products. Hence, the loss of those profits resulting from informing consumers is not unfair. (Streiffer and Rubel 2004: 240)

Rubel and Streiffer's argument here is:

1. Mandatory GE labels would deprive the marketers of GE products of some profits, but those profits are ones they would have received only through consumer ignorance of the fact that their products are GE.
2. Marketers are not entitled to any profits gained through consumer ignorance.
3. The profits that the marketers of GE products lose due to mandatory GE labels are not profits to which the marketers would have been entitled.
4. If the profits lost are not ones to which the marketers of GE products would have been entitled, then their loss of those profits is not unfair.
5. Therefore, mandatory GE labels would deprive the marketers of GE products of some profits, but their loss of those profits would not be unfair.

The first problem with the argument is that it addresses only one of several concerns about the fairness of mandatory GE labels. It focuses on the financial cost of mandatory labels for the marketers of GE products. Other issues of fairness range from the loss of liberty for both the marketers who will no longer be able to sell nonlabeled GE products and the consumers who will no longer be able to purchase them to the increased financial cost to those consumers who do not value the information provided by GE labels to the point where they are willing to accept an increased purchase cost to obtain it. The second problem with argument is that its first premise is unsupported. Why should we assume that the financial cost to marketers of a mandatory GE labeling scheme is equal to, or perhaps less than, the amount of profits they lose because some consumers, who would have purchased their products in ignorance of the fact that they are GE, refuse to do so once they learn that fact? Rubel and Streiffer offer no answer. The third problem with the argument is that its second premise is clearly false. Marketers are often entitled to profits they gain through consumer ignorance.

Consider again the example of the car dealer who does not mislead us into thinking the car we are purchasing will not be on sale in the near future but also does not correct our false belief that it will not be. If we had known of the upcoming sale, we would have delayed our purchase and saved, say, a thousand dollars. His profits would have been less by just that amount. He thus gains some of his profit due to our ignorance. He is, however, fully entitled to it. If he is not, it is perfectly appropriate for us to demand it back, which it clearly is not. Rubel and Streiffer need to revise their second premise so that it somehow does not apply to the sale of some products, for example, automobiles, but does apply to the sale of GE products. To do this, however, they have to find a relevant difference between the exchanges, and while there are some to which they might appeal, for example, the supposed increased risk associated with GE, as opposed to non-GE, products, expanding their argument in this way will take them beyond considerations of consumer autonomy. They will have to confront the difficult issues the autonomy argument was supposed to avoid.

In all, the democratic equality argument joins the other versions of the autonomy argument in failing to establish a moral obligation for the government to legally mandate GE labels.

## CONCLUSION

I have examined three versions of the autonomy argument. In each case, examination has shown that the argument is unsuccessful in fairly obvious ways. A successful version may be out there somewhere, but it is highly unlikely. The primary problem facing the autonomy argument is that the promotion of consumer autonomy is not sufficient, in itself, to obligate the government to require labels. Any one of a wide variety of labels will promote consumer autonomy, from labels listing competitors' prices for the same item, to country of origin labels, to "environment friendly" labels, to labels announcing upcoming sales. Certainly, the government is not obligated to require them all. So what makes the case of GE labels different?

There are plausible answers. Perhaps, GE products carry a greater risk to consumers than their non-GE counterparts. Perhaps, society has a special interest in promoting GE-based purchase decisions just as it has a special interest in promoting health-based purchase decisions or energy efficiency-based purchase decisions. Yet, if it is developed to rely on such points as these, the autonomy argument loses its most attractive feature. It is no longer independent of claims about the nature, risks and benefits of GE products.

None of this should surprise us. Proposals that promise to solve difficult problems without confronting the issues that make them difficult are always too good to be true. Proponents of the autonomy argument tell us we can establish the obligatory nature of mandatory GE labels without wrestling with the risk and safety of GE products or carefully detailing and weighing the conflicting interests of various stakeholders, or even weighing in the balance the general social value of promoting consumer awareness of GE products. We need only note that a significant number of consumers value having the information and the conclusion that GE labels should be mandatory will follow. Yet, of course, it does not. It is time to put the autonomy argument aside and confront the real issues.

## Notes

Paul Weirich provided helpful comments on an earlier draft of this chapter. I have gained much from discussing the issues with Alan Rubel and Robert Streiffer, despite our continuing disagreements.

1. Some take the autonomy argument to be more fundamental from an ethical perspective than mandatory labeling justifications based on risk or safety. Consider Jackson (2000: 325–326):

> Labeling on foods, clothing, medicines and other products can have two purposes. One is to help consumers manage risk. The other is to protect the individual autonomy of consumers to choose which products they wish to purchase. Arguably, the latter of these two purposes is most fundamental from an ethical standpoint; for, although total paternalism (i.e., where the experts make all the choices) might well turn out to be optimal risk management, no one would argue that efficient risk management would justify such a blatant challenge to autonomy.

2. Hansen (2004) ultimately rejects the autonomy argument.
3. In focusing on the first stage of the argument, I do not suggest that the second stage is without problems or interesting issues. Its success clearly depends on the resolution of such difficult and controversial issues as the appropriate limits of government regulation of commercial activity and the ability of market forces to support the ethical behavior of marketers. I focus on the first stage for reasons of space and because the problems there have been frequently ignored.
4. While the number of consumers whose autonomy is promoted by GE labels is not relevant for this stage of the argument, it may well be relevant in the attempt to infer the mandatory labeling thesis from the marketers' responsibility thesis. A government's obligation to legally require GE labels may depend, in part, on the number of people who are served by ensuring that those marketing GE products meet their obligation to label.
5. Some may wish to question the premise that prima facie duties become all-things-considered obligatory unless there is a morally sufficient reason not to perform them. Why not adopt an alternative view, that prima facie duties are morally permissible, but perhaps not obligatory, if we lack a morally sufficient reason not to perform them?
6. As even proponents of mandatory labeling admit (Streiffer and Rubel 2004: 232), information on the GE nature of most products is already readily available to consumers who are willing to seek it out.
7. Streiffer and Rubel have suggested this defense of the autonomy argument in conversation.
8. The marketers of GE products do represent them as being safe and often as being materially the same as non-GE alternatives. Some may question whether these claims are unethically deceptive, but then the case for a marketer's moral obligation to label GE products will rest on premises regarding product safety and the material difference between GE products and non-GE alternatives. The autonomy argument is supposed to avoid such claims.
9. Note, however, that the second objection to the first version of the autonomy argument applies here also. To establish their second premise that the FDA will promote consumer autonomy by mandating GE labels, Rubel and Streiffer must show more than that consumers desire the information. They must show that having the information will lead consumers to make product choices that are more in line with their values.
10. For an analogous case, consider why the government requires energy-efficiency labels on major appliances. It is not simply that consumers want the information and the government is obligated to promote consumer autonomy. It is that society has an interest in promoting energy efficiency-based purchase decisions.

## References

Hansen, Kirsten. 2004. Does Autonomy Count in Favor of Labeling Genetically Modified Food? *Journal of Agricultural and Environmental Ethics* 17: 67–76.

Jackson, Debra. 2000. Labeling Products of Biotechnology: Toward Communication and Consent. *Journal of Agricultural and Environmental Ethics* 12: 319–330.

Rubel, Alan, and Streiffer, Robert. 2005. Respecting the Autonomy of European and American Consumers: Defending Positive Labels on GM Foods. *Journal of Agricultural and Environmental Ethics* 18: 75–84.

Streiffer, Robert, and Rubel, Alan. 2004. Democratic Principles and Mandatory Labeling of Genetically Engineered Food. *Public Affairs Quarterly* 18: 223–248.

# 7

# Consumer Response to Mandated Labeling of Genetically Modified Foods

*Nicholas Kalaitzandonakes,*
*Leonie A. Marks & Steven S. Vickner*

Numerous studies have elicited consumer attitudes and stated preferences toward genetically modified (GM) foods and mandatory GM food labels. These studies indicate that consumers often seem ambivalent, if not negative toward, such foods (Select Committee on Environmental Audit 2003; Marris et al. 2001; Petts et al. 2001; European Commission 1991, 1993, 1997, 2000, 2003; PIFB 2001, 2003, 2004, 2005; Hornig Priest 2000). Furthermore, it has generally been assumed that elicited attitudes and stated preferences are reflective of how consumers would respond when confronted with a choice between GM and non-GM foods. Along these lines, elicited consumer attitudes and stated preferences have been used to support the case for mandatory labeling of GM foods in order to protect the consumer "right to know" and "consumer autonomy" (e.g., Streiffer and Rubel 2003; Rubel and Streiffer 2005).

However, as we argue in this chapter, attitudes and stated preferences are not the only, or even the best, way to gauge actual consumer preferences toward and demand for GM and non-GM foods. And while attitudes and stated preferences might be ideal for such assessments prior to the commercialization of GM foods, revealed preferences should be given more attention after market introduction. With 10 years of biotech crop and GM food production, it is surprising that little attention has been given to market evidence of how consumers have actually responded to GM foods, especially those carrying mandated GM labels.

In this chapter, we examine the bases for the theoretical and empirical divide between stated and revealed preferences. We also examine stated and revealed preferences toward labeled GM foods in two separate countries. Our empirical results suggest a divergence between the two sets of preferences and call into question sole reliance on attitude and stated preference studies, as well as the usual conventional wisdom of broad consumer rejection of GM foods in the face of choice.

## LITERATURE REVIEW

A key methodological problem with many prior studies gauging consumer preferences and response to GM foods arises from the fact that consumer choice has been typically reduced to consumer "acceptance." While rarely defined, "acceptance" seems to be a catchall term for everything that has to do with public sentiments toward GM foods. As Durant et al. (1998: 9) argue, however, "public acceptance" as a concept is both semantically and pragmatically flawed. It implies that the public has only a discrete choice to make (accept or reject). Yet such "lexicographic" consumer preferences are rare. In most cases, consumer choices are rich and variable, responsive to changes in prices, income, information, social norms, and a variety of other factors.

Durant et al. (1998) prefer to measure "public perceptions" instead—or the ideas, interest in, understanding of, and attitudes toward biotechnology that the public holds. But the concept of "consumer perceptions" can vary from one study to another. For example, Pennings et al. (2002) define perception as a "consumer's interpretation of the chance to be exposed to the content of a risk and a consumer's assessment of the risk inherent in a particular situation." Pennings et al. reserve the term "attitude" to describe "a consumer's general predisposition to [the risk]." While consumer attitudes and perceptions can be defined somewhat differently from one study to another, empirical measures of public perceptions and attitudes can be even more variable.

### Empirical Evidence from Attitudinal Surveys on GM Foods

Attitude surveys about GM foods range from the in-depth studies carried out by academics, government agencies, and various interest groups (see, e.g., Durant et al. 1998; European Commission 1991, 1993, 1997, 2000, 2003; Gaskell et al. 1999; Hoban 1996, 1998; Hornig Priest 2000; Center for Science in the Public Interest 2001; IFIC 2001a, 2001b, 2002, 2003, 2005; PIFB 2001, 2003, 2004, 2005) to up-to-the-minute polls carried out by mass media and other organizations (see, e.g., MSNBC 2000; Walsh 1999). A wide range of questions has been asked. Importantly, responses vary considerably depending on how questions are posed and the kind of sample used (e.g., size, demographics, location), and vary over time.

The Pew Initiative on Food Biotech (PIFB 2001, 2003, 2004, 2005) has tracked U.S. public attitudes toward GM foods over the last five years. So has the International Food Information Council (IFIC), and their respective conclusions seem at odds. The 2001 PIFB survey concluded that "most Americans (58%) opposed the introduction of GM foods into the food supply," while its October 2005 survey found that "50% of consumers opposed their introduction" (PIFB 2005: 6). Yet IFIC, based on its September 2001 survey, concluded that "65% of Americans were likely to buy genetically modified produce."[1] Similarly, the latest survey by IFIC (2005) found a majority of Americans (64%) likely to purchase modified produce.

Empirical measures of perceived risks associated with GM foods have also varied drastically depending on how relevant survey questions have been asked.

Unprompted, 5–11% of U.K. survey respondents listed biotech foods as a food safety concern (Food Standards Agency 2001, 2002). Moreover, in the 2001 survey, when asked, "Which foods do you no longer eat nowadays because of food safety concerns?" only 1% of U.K. respondents mentioned food with GM ingredients (Food Standards Agency 2001). Similar results have been obtained in the United States, where up to 1% percent of U.S. unprompted respondents mentioned GM foods as a potential food safety hazard (IFIC 2005). However, when prompted, anywhere from 16% to 34% of U.S. respondents (PIFB 2001) and from 21% to 38% of U.K. respondents (Food Standards Agency 2002, 2004) stated they were concerned.

Empirical measures of consumer attitudes toward mandatory labeling of GM foods are subject to the same kind of variation. In the IFIC (2001a) survey, when respondents were presented with a question stating that biotechnology critics demanded mandatory labeling of all foods produced through biotechnology, even if the safety and nutritional content of those foods are unchanged, more than 50% of the respondents sided with the critics. This result was inconsistent with that of a preceding question in the same survey that asked the respondents whether they supported the current FDA policy of voluntary labeling—70% said they supported the policy. Unprompted, only 1–2% of the respondents in the IFIC's more recent survey "could think of any information that is not currently included on foods labels that [they] would like to see on food labels" (IFIC 2005). This low level of concern has remained constant since first tracked by the IFIC in 2001.

In contrast, other surveys have arrived at exactly the opposite conclusion. In 2004, the PIFB concluded that a high percentage of American respondents (80%) strongly favor "labeling all food that is genetically modified" (PIFB 2004). As early as 1992, Hoban and Kendall (1993) found 85% of U.S. consumers polled thought labels were very important. And such high levels of public demand for mandatory labels are also reflected in international polls conducted at various points in time. For instance, several studies concluded that there was significant interest among European consumers for mandatory labeling of GM foods just as relevant regulations were being developed by the European Commission. The Eurobarometer found that 74% of respondents wanted special labels on GM foods. A poll of 1,518 U.K. respondents by the *London Evening Standard* found 78% of them wanted GM foods clearly labeled in 1999 (Fletcher 1999). An opinion poll conducted by the *Guardian* and ICM Research in 1998 found that 96% of 500 respondents indicated that they wanted mandatory GM food labels (Vidal 1998).

## Attitudes as Predictors of Consumer Preferences and Behavior

Divergent empirical measures of consumer perceptions and attitudes toward GM foods make inferences about the underlying preferences of consumers difficult. Importantly, there are also good theoretical reasons as to why elicited consumer attitudes and actual consumer preferences toward GM foods could differ. Indeed, it is well documented in the literature that sentiments expressed in attitudinal surveys are often poor predictors of preferences and behavior (Deutscher 1973; McFadden 1999; V. L. Smith 1991; Kahneman et al. 1999; La Piere 1935).

Attitude surveys can place consumers in "symbolic situations" that do not require any follow-up action and may not equate with actual decision contexts that change from one time to the next (Deutscher 1973; V. L. Smith 1991; La Piere 1935). Furthermore, such surveys do not distinguish between an individual's role as a consumer from their role as a citizen (Nyborg 2000). In the role of citizen, individuals may be more focused on public interest. In their role as consumers, they may more focused on their personal goals (Sagoff 1988: 8). Every individual may have distinct and possibly conflicting preference orderings over social states. Choices made (or stated) in one context may not be replicated in another (Nyborg 2000).

Cognitive biases can also play an important role on how information is received and processed by survey respondents (Kahneman et al. 1999). How questions are framed, the order in which information is presented, the degree of knowledge, and the understanding of the respondent are all potential sources of bias and error (Kahneman and Tversky 1984; Tolley and Randall 1983). Cognitive biases can thus significantly obscure the relationship between elicited attitudes and actual consumer preferences and behavior (e.g., Ajzen and Fishbein 1977).

## Experimental Approaches as Predictors of Consumer Preferences and Behavior

Probably the most important limitation of attitude surveys is that they do not account for potential price and income effects. Yet consumer behavior is typically driven by such economic factors. Choice and auction market experimental designs elicit consumer preferences and expected behavior by studying responses to hypothetical price changes across different income groups.

### Willingness-to-Pay Studies

Theoretically, consumers should be willing to pay more for GM foods exhibiting desirable attributes (e.g., foods with improved organoleptic properties) (Boccaletti and Morro 2000). Alternatively, some consumers might be willing to pay more to avoid them altogether (Moon and Balasubramanian 2001). Contingent valuation is the best-known, most frequently used willingness-to-pay method.

Empirical willingness-to-pay studies suggest that a sizable portion of survey respondents would generally prefer non-GM over GM foods if faced with the choice. For example, Chern et al. (2002), in a survey of Norwegian and U.S. consumers, found that 80% of the Norwegian respondents and more than 60% of U.S. respondents would chose non-GM and non-GM-fed salmon over salmon that was itself GM or was fed with GM feeds. Lusk et al. (2003) evaluated consumer purchasing intentions toward beef ribeye steaks in France, Germany, the United Kingdom, and the United States and found that survey respondents in these countries indicated that they would be willing to pay a significant premium (US$7.29 to US$9.94 per lb) for nonhormone beef and anywhere from US$3.31 to US$9.32 for non-GM-fed beef. Burton and Pearse (2002) examined purchase intentions toward beer products in Australia. They found that 30% of the respondents stated that they would avoid GM beer, and 70% indicated that they would purchase it at some discount. Moon and Balasubramanian (2001) found that 44% of U.S. and 71% of U.K. survey respondents preferred non-GM corn flakes. According to their study, only 6% of U.S. and 2% of U.K. survey respondents preferred GM foods.

While informative, consumer stated preferences elicited from willingness-to-pay studies are not without limitations. A common criticism is that willingness-to-pay measures result from hypothetical questions (Tolley and Randall 1983; Prato 1998). This criticism is often stated as follows: If you ask a hypothetical question, you get a hypothetical answer. Contingent valuation methods run the risk of giving unreliable results particularly when respondents are not well enough informed about the subject to state their willingness to pay or understand the question being asked. This criticism is particularly pertinent to GM food studies, where consumers can exhibit incomplete knowledge about the technology.

The approach is also susceptible to other important types of bias—strategic bias, hypothetical bias, and anchoring effects, among others. Strategic bias can occur when consumers deliberately understate or overstate the true value they place on an attribute—for example, if they believe that by so doing they might influence a policy outcome. Hypothetical bias, on the other hand, typically occurs when consumers are unable to accurately assess their willingness to pay. Hypothetical bias is possible even in well-designed surveys, particularly when consumers have limited prior experience with the attribute (in this case, GM vs. non-GM food products). Lack of actual purchasing experience can make it very difficult for consumers to become aware of their own preferences so that they can place a value on changes in price, quantity, and quality (Prato 1998). Anchoring effects occur when consumers are induced to consider dollar amounts as candidate answers—their response can be strongly biased toward any value, even if it is arbitrary (Kahneman et al. 1999). In all such instances, one might expect to observe discrepancies between stated and actual consumer preferences or behavior.

### Experimental Auction Studies

In order to address some of the limitations inherent in contingent valuation and attitude surveys, several recent studies have used experimental auction markets to elicit purchase intentions toward GM foods (Rousu et al. 2003, 2004; Huffman et al. 2001, 2002, 2003; Lusk et al. 2004; Noussair et al. 2004; Van Wechel et al. 2003). Experimental auction markets are usually run in a lab environment that allows researchers to collect detailed information about the participants in the experiment, to introduce different types of "information shocks," and to observe changes in participants' behavior (Shogren et al. 1999). Participants are often asked to bid for different goods, paying "real" money for the purchase of "real" goods at the end of the experiment. Such bidding behavior can be more revealing of consumer preferences and purchasing intentions.

Experimental retail and auction markets provide a more realistic environment for eliciting consumer preferences. And economists have developed propositions from theory and experimental analyses that predict which auctions will yield a stable market price and provide incentives for consumers to bid their true reservation value (Menkhaus et al. 1992). Likewise, experimental auction markets allow researchers to ask attitudinal questions of consumers. Within this context, they can monitor the impact of information on consumers to determine how their attitudes change (or not) as a result.

It is interesting to note that experimental auction market studies have arrived at somewhat different conclusions from those derived from attitude surveys and choice experiments. For instance, only a minority of French respondents have indicated that they were supportive of GM foods (54% were supportive in 1996, 35% in 1999, and 30% in 2002) in attitude surveys and willingness-to-pay studies (European Commission 2003: 18). However, in an experimental laboratory setting, Noussair et al. (2004) found that a smaller minority (35%) of French consumers who participated in the experiment initially boycotted GM-labeled biscuits after learning that they contained GM ingredients, while 65% of these consumers were willing to purchase them. Specifically, 40% were willing to purchase them at some price (discount), and 25% of the participants were indifferent and would purchase regardless. Noussair et al. (2004) also found that subjects who had previously discounted their bids for a selected GM food item were reassured by the brand identity after the brand of the GM-labeled product was revealed. It is also interesting to note that the participating consumers were asked to complete a survey prior to conducting the experiments. In this prior survey, 91% of them indicated that they would not purchase foods containing GM ingredients.

Similar diverging conclusions have been found in the United States. Studies based on attitude and willingness to pay have concluded that anywhere from 14% to 69% of U.S. consumers would not purchase GM foods. However, experimental auction market studies found that only 13–24% of U.S. consumers who participated in such experiments were unwilling to purchase GM foods (Van Wechel et al. 2003; Buhr et al. 1993). Indeed, Buhr et al. (1993) found that 87% of U.S. consumers were willing to pay a premium for leaner GM pork with fewer calories produced using porcine somatotropin. This result contrasted with earlier attitude surveys indicating that consumers would avoid such products (Hoban and Burkhardt 1991).

There are, however, some limitations to experimental auction market approaches, as well. The range of items for purchase is much more limited than in an actual retail store. In addition, participants may fall foul of what is known as the "Hawthorne effect," where respondents inflate the bids to please the monitor of the experiment (Shogren et al. 1999: 1192). In addition, such approaches cannot be readily applied to a random sample of the population (Cropper 1995) because they are geographically (and therefore demographically) limited in scope. Accordingly, the results of such studies are not directly generalizable.

Of course, opinion polls, attitude surveys, choice experiments, and experimental auction markets all attempt to forecast what consumers would do when confronted with relevant choices between GM and non-GM food items. However, since GM foods have been in the marketplace for a number of years, one need not forecast consumer preferences and behavior but may rather directly observe them. Surprisingly, there is not a single study of revealed consumer preferences toward GM foods. Informal market evidence would indicate that broad consumer avoidance of foods with GM ingredients or meat products from livestock fed with GM ingredients has not been observed in any country. Yet, a case can be made that, in the absence of GM food labels, consumers might lack sufficient information to properly reveal their preferences.[2] Instead, labeled GM foods offered right next to non-GM foods provide a more proper context for studying revealed consumer preferences. Along these lines, we present here results of our study of consumer behavior toward foods labeled for GM content over an extended period of time in supermarkets in the Netherlands and China.

## MARKET EVIDENCE OF CONSUMER RESPONSE

There is some limited anecdotal evidence of how consumers have responded when confronted with GM-labeled foods. The first case occurred in the United Kingdom from 1996 to 1999. Prior to its removal from U.K. store shelves in 1999, a GM tomato puree (labeled as such) was sold side by side with the non-GM products, at the same price, although the GM variety was in a 10% bigger can to pass on the process cost savings to consumers. The cans outsold competing non-GM brands by a significant margin (Nunn 2000). Other evidence can also be found from Japan. Executives of the Japanese supermarket chain Jusco reported that after one year of offering a number of labeled GM foods in their stores, no significant impact on their sales could be detected (Hur 2001)—this despite public opinion and attitude surveys consistently finding a majority of Japanese consumers to be negatively predisposed toward GM foods. While such evidence is anecdotal and information is sketchy, the implied consumer behavior and preferences are opposite to those anticipated from the numerous stated preference studies. Here, we offer new and more formal empirical evidence of revealed consumer preferences toward GM-labeled foods, first in the Netherlands and then in China.

### Consumer Behavior toward Mandated Food Labels in the Netherlands

The Netherlands is a small, highly industrialized, high-income nation within the European Union. Its current population is approximately 16 million. It has a net trade surplus and is an important exporter of food.

The first biotech food product to be considered for marketing in the Netherlands was the growth hormone bovine somatotropin (BST) in 1987. However, BST was never allowed in the Dutch market because of the E.U. moratorium that came into effect in 1990. The next product to be marketed was the recombinant DNA cheese rennet chymosin, which was approved in 1990. This was followed by the arrival of GM soybeans in 1996, approved by the Dutch Ministry of Health. In April 1997, the ministry issued, within the framework of the 2002 E.U. Food Law, a labeling directive on the use of GM soybeans and maize products (Midden et al. 1998). However, in October 1997 this ruling was revoked by the courts in compliance with E.U. regulations, and from November 1997 onward the labeling of GM soybeans was directed by the European Union.

Because of the early court rulings, the Netherlands was one of the first E.U. countries to enact mandatory labeling. A number of food products sold in Dutch supermarkets at that time had to be labeled because they contained GM soy ingredients. Either their ingredient labels listed *soja* (soya) that was asterisked with an explanatory label underneath stating *genetisch gemodificeerd* (GM), or a general statement was included under the ingredients list as follows: *geoproduceerd met genetisch gemodificeerde soja*, that is, "made from genetically modified soya." We estimate that about 200 prepared and processed foods carried such labels in the Dutch market during the late 1990s.

How did Dutch consumers feel about GM foods at that time? How well informed were they about biotechnology and the availability of GM foods in the marketplace? And, importantly, what did they do when they were confronted with a choice at the supermarket shelves?

## Public Understanding of Biotechnology and Attitudes toward GM Foods

Over the years, Dutch citizens have seemed well informed about biotechnology and GM foods (Hoban 1997; Hamstra 1993; Hamstra and Smink 1996). The Dutch have consistently scored higher than the European average in terms of their technical knowledge about biotechnology (6.25 on a scale of 1 to 10, as opposed to the European average of 5.46) (European Commission 2003). Furthermore, the Dutch are generally considered informed food consumers as the Netherlands has had a long history of active consumer involvement in food and nutrition policy. More than 10% of Dutch households belong to consumer organizations (Hillers and Lowik 1998).

While Dutch consumers have been consistently informed about biotechnology and GM foods, their attitudes toward such foods have varied over time. In 1996, 56% of Dutch respondents indicated that GM foods should be encouraged (European Commission 1997). Much like the rest of Europe, however, only 36% of Dutch consumers responding to surveys indicated that they "would buy genetically modified foods" at that time (European Commission 1997: 52). The number of Dutch consumers indicating in relevant surveys that they would purchase GM foods dropped further to 30% by 1999 (European Commission 2000: 60) and to 28% by 2002 (European Commission 2003).[3]

Temporal changes in attitudes could be related to changes in the amount and tone of information on GM foods received by Dutch consumers. Information about biotechnology and GM foods is typically event driven and tends to vary over time. Spikes in coverage around events and technological controversies can raise public awareness about an issue and, in turn, affect perceptions. Studies have also found a link between food safety information and consumer purchasing behavior (van Ravenswaay and Hoehn 1991; Dahlgran and Fairchild 2002; Wessels et al. 1995; Kalaitzandonakes 2004). It is therefore important to look at the kind of information Dutch consumers were exposed to during 1997–2002.

Most Dutch consumers get their information about biotechnology from the media. The average person in the Netherlands spends half an hour a day reading at least one newspaper (Midden et al. 1998: 105). Ninety-two percent of the

Netherlands' public thinks that newspapers do good work for society (European Commission 2003).

Dutch presses can be distinguished by the type of information that they publish. There are popular general interest papers, such as *Telegraaf* and *Algemeen Dagblad*, that emphasize human interest stories, entertainment, and sports. The quality press includes papers such as *De Volkskrant*, which are more policy oriented, provide science coverage, and cater to middle- to upper-class, more educated readers (Midden et al. 1998). There is also the Algemeen Nederlands Persbureau, which is a national press agency, similar to the Associated Press, and which provides newspapers, media, and television with domestic and international news.

In order to examine the degree to which Dutch consumers were exposed to coverage of issues related to GM foods in the late 1990s and early 2000, we measured the frequency of such coverage by the Algemeen Nederlands Persbureau. Article frequency is indicative of the salience of an issue or the "news value" of a particular development. It is also indicative of the intensity of media attention. Some mass communication scholars argue that as media coverage intensifies, so public opinion on an issue becomes more negative (regardless of the tone of such coverage) (Leahy and Mazur 1980).

Figure 7.1 details our findings. Consistent with the earlier work of Midden et al. (1998), who examined media coverage of biotechnology from 1972 through 1996, we also found that the salience of the issue of GM foods was high at the end of the 1990s—peaking in 2000. Hence, during the period of interest when GM labels were in place (1997–2000), Dutch consumers were exposed to increasingly intense media coverage on the issue.

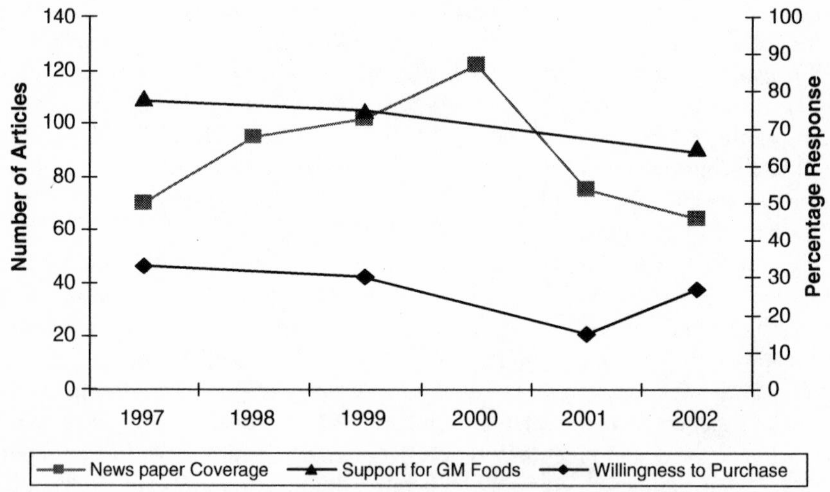


**Figure 7.1.** Salience, willingness to purchase, and support for GM foods in the Netherlands, 1997–2002. Note: data on willingness to purchase and support are from the Eurobarometer surveys (European Commission, 1997, 2000, 2003). The 2001 willingness to purchase figure is based on a survey conducted by the NIPO Research Bureau.

Based on the results of the various attitude surveys carried out during the period of interest, one would expect a sizable market segment (anywhere from 44% to 85% of Dutch consumers) to avoid GM foods if confronted with the choice. Likewise, one would expect Dutch consumers to be armed with information about GM foods based on their high level of awareness, knowledge, and intense media coverage. Dutch consumers were indeed confronted with such choice for a meaningful number of processed foods. The question, then, is how did they *actually* chose?

## Market Evidence of Consumer Response to GM Labels in the Netherlands

In order to examine consumer response to GM foods in the Netherlands, we used national-level, syndicated point-of-purchase grocery store scanner data. The data were collected from supermarkets with sales exceeding $2 million across the Netherlands. The data set was acquired from a commercial marketing research vendor. Four product categories are included in the data set: frozen processed meat, frozen pizza, frozen processed fish, and canned soup. The data set spans 260 consecutive weeks, from the week ending Sunday, April 27, 1997, to the week ending Sunday, April 14, 2002, and includes products that contained GM ingredients as well as products that did not. Across those relevant products containing GM soybean ingredients, labels were introduced in the ninth week, the week ending Sunday, June 22, 1997. The labels remained on relevant products for 151 consecutive weeks and were removed the week ending Sunday, May 14, 2000, as non-GM ingredients were sourced by food manufacturers.

Figures 7.2–7.5 illustrate the sales (in total equivalent units) for the processed foods that carried positive GM labels during 1997–2000. From these illustrations, it is easy to see that, in aggregate, the Dutch consumers did not significantly

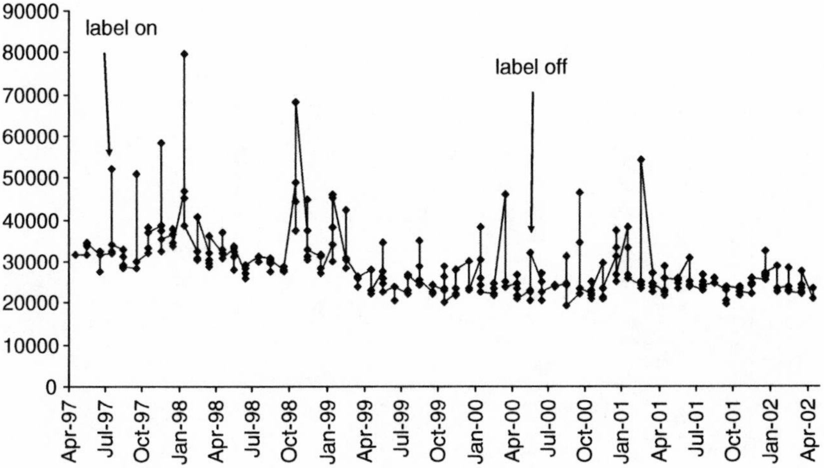


**Figure 7.2.** Quantity of GM-labeled soup sold.

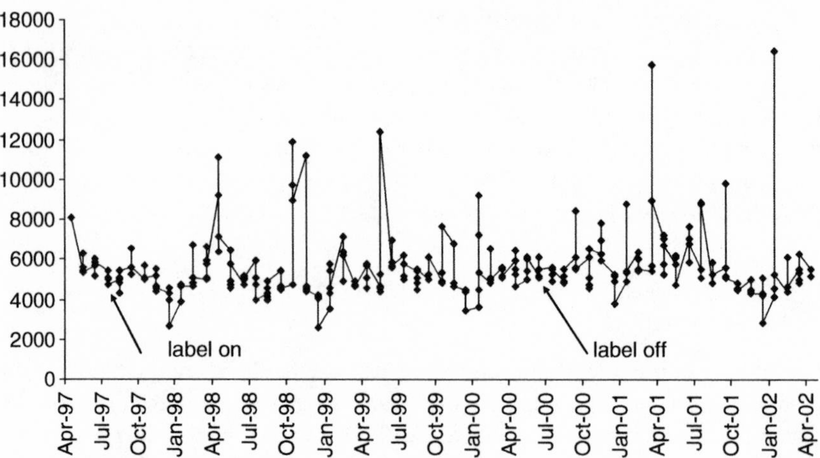


**Figure 7.3.** Quantity of GM-labeled frozen fish sold.

change their purchasing behavior toward foods that received labels indicating the presence of GM ingredients. Nor did they alter their purchasing behavior toward such foods after the labels were removed nearly three years later. There are no abrupt adjustments or gradual shifts over time away from GM foods. Of course, in order to strengthen these conclusions, one should explicitly test consumer behavior within the context of a statistical model that incorporates the role of information (see Wessels et al. 1995; M. E. Smith et al. 1988; Swartz and Strand 1981; Piggott and Marsh 2004), changes in relative prices, real per capita expenditure, seasonality, and any other relevant factors. Given the focus of this chapter, we do not expand on the technicalities of the relevant econometric model here;[4] however, statistical analysis of our data does support the conclusion that

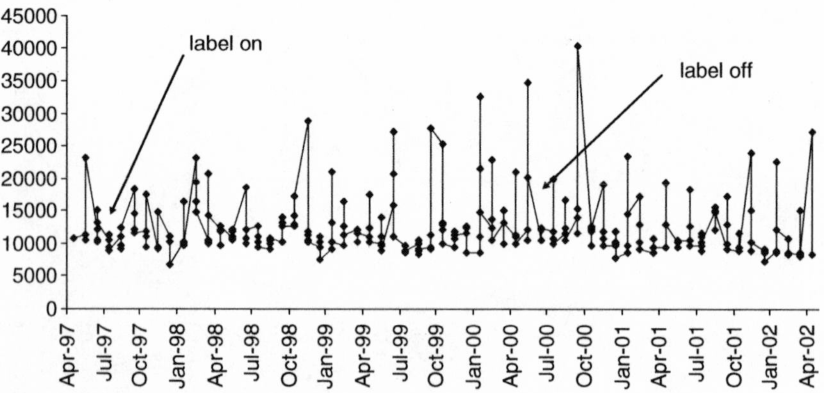


**Figure 7.4.** Quantity of GM-labeled frozen pizza sold.

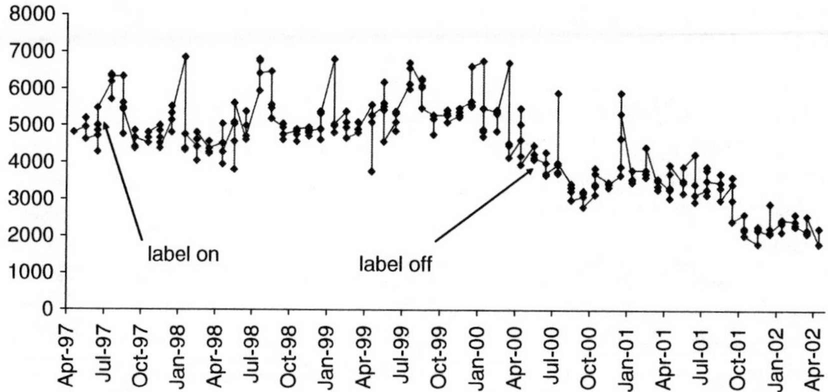


**Figure 7.5.** Quantity of GM-labeled frozen meatballs sold.

consumers did not "vote with their wallets" when they had the opportunity to do so—both when the labels went on and when they came off.

## Consumer Behavior toward Mandated Food Labels in China

China is an emerging economy, with the largest population in the world. In 1999, 1.306 billion people resided in China—about 21% of the world total (Information Office of the State Council 2000). In 2003, total consumer spending in China was US$1.1 trillion, of which roughly 41% was spent on food (Deloitte Development LLC 2005: 2). From 1997 to 2003, per capita disposable income grew by 64% while per capita spending of urban consumers rose 60% (Deloitte Development LLC 2005: 2). Similarly, inflation-adjusted per capita consumer expenditures on food increased by more than 70% from 1994 to 2003 (Gale 2005).

China is also the fourth largest producer of biotech crops after the United States, Argentina, and Canada (Ho and Vermeer 2004) and has a sizable biotech research and development infrastructure already in place (Huang et al. 2002). In July 2001, the Ministry of Agriculture introduced "Regulations on Labeling Agricultural Transgenic Organisms" (Ho and Vermeer 2004), and on March 20, 2002, it implemented mandatory labels on all foods sold in supermarkets containing ingredients from GM soybeans (seeds, oil, soy powder, soybeans, soybean dregs), corn (seeds, corn oil, corn powder), rapeseed (rapeseed dregs), tomatoes (seed, tomato, tomato sauce), and cottonseed (Li et al. 2002). However, it was not until July 2003 that these regulations began to be implemented domestically—and to date, soybean oil is the main product category that has carried mandatory GM labels.

How do Chinese consumers feel about GM foods and mandatory labels? How well informed are they about biotechnology and the availability of GM foods in the marketplace? And, importantly, what do they do when they are confronted with a choice at the supermarket shelves?

### Public Knowledge of Biotechnology and Attitudes toward GM Foods in China

A handful of studies have evaluated Chinese consumer awareness of GM foods and attitudes toward them. Awareness and knowledge about GM foods among respondents in relevant surveys have varied depending on the sample and the type of question asked, and have varied over time. Hu and Chen (2004) investigated consumer purchase intentions of GM vegetable oil. Their survey was conducted in winter 2002 and spring 2003. They concluded that respondent knowledge and awareness about GM technology was limited. Approximately 40% of respondents did not know what GM stands for, and 60% were unsure whether vegetable oil made from GM oilseed was sold on the market (while it was probably sold at the time, it was not labeled). Similarly, a Greenpeace survey conducted in January 2003 of 1,000 Guangzhou respondents found that 64% did not know whether their supermarkets sold GM food products (Greenpeace International 2003). On the other hand, 71% of Chinese respondents indicated that they had "heard of transgenic (GM) food products," in a survey conducted by Ho and Vermeer (2004) in September 2003. Among higher educated consumers, 82% had heard of biotech food products. Their study was conducted two months after GM food labels had been in place. The authors of the study concluded that the higher awareness compared to earlier surveys was the result of government campaigns to enforce labeling of GM foods in Chinese supermarkets. In July 2003, manufacturers of 14 brands of soybean oil were fined for not labeling GM ingredients (Ho and Vermeer 2004: 161), an event that was broadly covered in mass media.

Chinese consumer attitudes toward GM foods and willingness to pay for them elicited through surveys have been similarly variable. For example, in their survey, Hu and Chen (2004) asked respondents, "Would you continue to buy an oil product if it was made from GM oilseeds?" Fifty-six percent indicated that they were unwilling to do so. Moreover, 44% indicated that they would choose non-GM over GM even if it cost 10% more. The survey by Ho and Vermeer (2004) asked, "Are you willing to consume food containing GM-based ingredients?" Forty-four percent of the respondents were willing to buy, while 51% were neutral.

From the above attitude and willingness-to-pay studies, one would expect that, initially, a large segment of Chinese consumers were unaware that a large share of the soybean oil sold in the market was of GM origin. One would expect, however, that consumer awareness of GM-labeled soybean oils improved through mass media coverage and, probably, over time through experience. Furthermore, these surveys would suggest that a sizable segment of Chinese consumers would avoid GM-labeled soybean oils and would favor other non-GM oils (e.g., non-GM soybean oil, rapeseed oil, peanut oil, corn oil, sunflower oil, and others) when confronted with the choice. So what did the Chinese consumers actually choose under these circumstances?

## Market Evidence of Consumer Response to GM Labels in China

In order to examine the response of Chinese consumers to GM-labeled soybean oils, we used national-level syndicated point of purchase data from a variety of retail outlets (e.g., hypermarkets, supermarkets, minimarkets, department stores, convenience stores, grocery stores, drug and other specialty stores).[5] The data set was provided by a commercial marketing research company, and it includes monthly volume and value sales for all major packaged edible oils sold in China.[6] The data set also includes detailed value and volume sales as well as price and market share information across various package sizes for selected branded soybean oils that have carried GM labels since July 2003.[7] The data set spans a total of 27 months, from January 2003 to the end of March 2005.

Figure 7.6 illustrates the volume and value shares of packaged soybean oils relative to all packaged edible oils sold in China over the period of interest. The figure illustrates that, as in the case of Dutch consumers, Chinese consumers, in aggregate, did not avoid GM-labeled soybean oils in favor of other nonlabeled oils, such as rapeseed, corn, peanut, and sesame oil, after the GM labels were introduced in July 2003. Indeed, after the introduction of GM labels, soybean oil gained share against all other major edible oils. Such gains are, in part, attributable to modest and gradual reductions in the relative price of soybean oil against other major edible oils (figure 7.6).[8]

Similarly, Figure 7.7 traces the aggregate value and volume shares of the top four national soybean oil brands, which account for 20–25% of all packaged edible oil sales and which have carried GM labels since July 2003. Figure 7.7 reinforces the point that Chinese consumers did not, in aggregate, discriminate against GM-labeled soybean oils. Indeed, the top four soybean oil brands have gained market share against most other edible packaged oils and have grown at similar rates with the rest of the packaged soybean oil brands sold in the Chinese market,

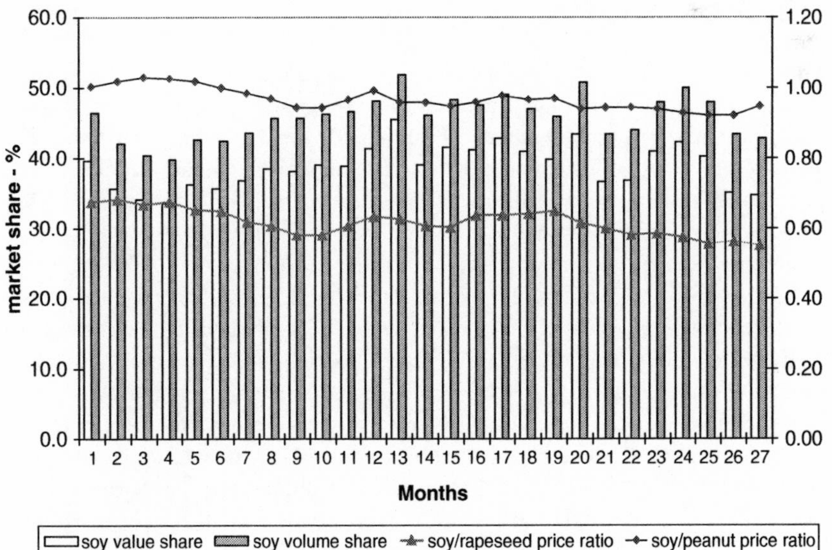


**Figure 7.6.** Value and volume shares of packaged soybean oils against total packaged edible oil sales in China, January 2003 through March 2005.

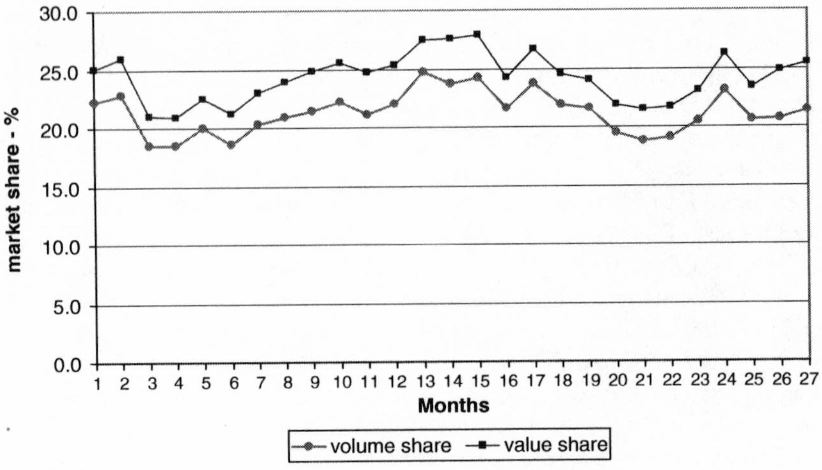


Figure 7.7. Value and volume shares of top four soybean oil brands against total packaged edible oil sales in China, January 2003 through March 2005.

many of which are derived from domestic conventional soybeans and have not carried GM labels. Hence, as in the Netherlands, we do not find evidence that Chinese consumers shifted their consumption away from GM-labeled foods while they had the choice to do so.

## IMPLICATIONS AND CONCLUDING COMMENTS

Hundreds of studies have elicited consumer stated preferences at various countries around the world. Such stated preferences have generally been viewed as proxies of potential responses toward GM foods. As we argue here, however, there are theoretical and methodological reasons as to why stated and actual consumer preferences and behavior could diverge. Unfortunately, little attention has been paid to the potential for such divergence, and little effort has been devoted to examining revealed consumer preferences toward GM foods.

In this study, we found empirical evidence of consumer revealed preferences toward branded food products carrying GM labels in two countries: the Netherlands and China. Our data sets are national in scope and cover multiple products—processed foods and vegetable oils. Our results suggest that at least in the Netherlands and China, a majority of consumers did not shift away from GM foods in the presence of alternatives. This finding stands in stark contrast to conclusions drawn from studies using attitude and willingness-to-pay surveys. We do not know why Dutch and Chinese consumers did not react to GM labels. It could be that consumers were reassured by trust in their food supply (Hamstra 1993) or by the brand identity of each of the labeled products (Noussair et al. 2004). Or, it could be that they were simply not concerned in the first place. In other words, it is possible that a majority of consumers read the labels, understood them, and kept on purchasing them regardless. Alternatively, it could be that Dutch and Chinese consumers did not read the labels (Noussair et al. 2004; Hu and Chen 2004; Greenpeace International 2003).

Irrespectively, the key result is that Dutch and Chinese consumers, in aggregate, did not change their behavior toward positively labeled foods with GM soybean ingredients or soybean oils over the study time periods. Much has been made of heterogeneous preferences and protecting consumers' "right to know," the "right to choose," and "autonomy" as the main reasons for mandatory GM food labels. In principle, there can be little objection to the argument that consumers need relevant information so that they can consistently act on their preferences. Market transparency is the linchpin of well-functioning markets. However, such arguments sidestep key issues such as the relevance of other instruments for avoidance of market failure (e.g., voluntary labeling) or the size of compliance costs and the concomitant efficiency of mandatory labels. Indeed, given that mandatory labeling systems are costly to implement (Kalaitzandonakes et al. 2001), costs and benefits associated with such labeling regimes must be carefully weighed in order to decide their optimality (Kalaitzandonakes 2004; Giannakas and Fulton 2002). Ultimately, the relevance of mandatory labels is an empirical question. In this context, the proportion of consumers that effectively discriminate between GM and conventional foods in the marketplace is a key parameter (Giannakas and Fulton 2002). Indeed, Caswell (1998, 2000) and Giannakas and Fulton (2002) have argued that a voluntary labeling program may better serve a country where only a minority of the population is interested in separating GM from non-GM foods. Our results, at a minimum, call into question the sole reliance on stated preferences to justify mandatory labeling regulation in any context. In the end, the potential benefits of mandatory labels need to be understood, documented, and weighed against regulatory costs. Understanding whether a majority of consumers, irrespective of motives, would use GM labels to discriminate against relevant products in markets is essential for effective policy especially in light of the credibility of labeling regimes, their high cost, and lack of direct evidence of market failure.

## Notes

This research was funded by the Illinois Missouri Biotechnology Alliance and the National Research Initiative of the Cooperative State Research Education and Extension Service, USDA, Grant #2005–35400–15413. Senior authorship is unassigned.

1. The respondents were asked whether they were "likely" or "unlikely" to buy a variety of produce, e.g., tomatoes or potatoes, if it had been modified by biotechnology to be protected from insect damage and require fewer pesticide applications.

2. An interesting case of diverging stated and revealed preferences played out when recombinant bovine somatotropin (rbST) was introduced in the United States in the mid-1990s. At the time of its introduction, various surveys indicated that almost 75% of U.S. consumers expressed concerns about the long-term health effects of consuming milk from dairy cows treated with the bioengineered hormone and intentions not to purchase it. After market introduction, however, a majority of U.S. consumers have continued to purchase

milk from rbST-treated cattle when offered side by side with "non-rbST" milk (Runge and Jackson 2000; Keisel et al. 2004). Indeed, today non-rbST milk sales make up a tiny fraction of the total U.S. milk market concentrated in certain regions of Wisconsin, Minnesota, California, and Vermont. Yet, in the absence of positive labels identifying the milk as sourced from cows treated with rbST, one could not directly evaluate the possibility of consumer avoidance.

3. It is worth noting that such measured attitudes placed Dutch consumers in the middle of Europe in terms of overall acceptance above countries such as France, Austria, Greece, and Italy but below countries such as Portugal, the United Kingdom, Spain, and Ireland.

4. The interested reader can find our theoretical approach and empirical results reported in Marks et al. (2004).

5. The data were collected in major cities and towns across China but not in villages. The market coverage of the sample has not been estimated by the data vendor.

6. Packaged edible oils represent an estimated 25% of all edible oils consumed in China. A large segment of Chinese consumers purchase edible oils in bulk. Packaged oil consumption is more common among urban consumers.

7. China is the single largest importer of soybeans in the world. A large share of such imported soybeans comes from the United States and Argentina, the two largest GM soybean producers in the world. Most major national brands of soybean oil, such as Fortune, Gold Ingots, Arawana, and King Elephant, are produced, at least in part, from imported GM soybeans, and accordingly, they are required to carry GM labels. Many smaller regional brands, especially those distributed in the northeast part of China, are produced from domestically produced, non-GM soybeans and are not required to carry GM labels. Hence, while not all the packaged soybean oil supply has been labeled, a large share (more than 65%) has.

8. While beyond the scope of this study, statistical analysis is under way to formally confirm this qualitative assessment.

## References

Ajzen, I., and M. Fishbein. 1977. Attitude-Behavior Relations: A Theoretical Analysis and Review of Empirical Research. *Psychological Bulletin,* 84(5) 888–918.

Boccaletti, S., and D. Morro. 2000. Consumer Willingness to Pay for GM Food Products in Italy. *AgBioForum* 3(4): 259–267.

Buhr, B.L., D.J. Hayes, J.F. Shogren, and J.B. Kliebenstein. 1993. Valuing Ambiguity: The Case of Genetically Engineered Growth Enhancers. *Journal of Agricultural and Resource Economics* 18(2): 175–184.

Burton, M., and D. Pearse. 2002. Consumer Attitudes towards Genetic Modification, Functional Foods and Microorganisms: A Choice Modeling Experiment for Beer. *AgBioForum* 5(2): 51–58.

Caswell, J.A. 1998. Should Use of Genetically Modified Organisms be Labeled? *AgBioForum* 1(1): 22–24.

Caswell, J.A. 2000. Labeling Policy for GMOs: To Each His Own? *AgBioForum* 3(1): 53–57.

Center for Science in the Public Interest. 2001. *National Opinion Poll on Labeling Genetically Modified Foods.* Available at www.cspinet.org/new/poll_gefoods.html.

Chern, W.S., K. Rickertsen, N. Tsuboi, and T.T. Fu. 2002. Consumer Acceptance and Willingness to Pay for Genetically Modified Vegetable Oil and Salmon: A Multiple-Country Assessment. *AgBioForum* 5(3): 105–112.

Cropper, M.L. 1995. Valuing Food Safety: Which Approaches to Use? In *Valuing Food Safety and Nutrition*, ed. J.S. Caswell. Boulder, Colo.: Westview Press, 207–210.

Dahlgran, R.A., and D.G. Fairchild. 2002. The Demand Impacts of Chicken Contamination Publicity—A Case Study. *Agribusiness* 18(4): 459–474.

Deloitte Development LLC. 2005. *China's Consumer Market: Opportunities and Risks.* Available at www.deloitte.com/dtt/cda/doc/content/DTT_DR_China_Consumer_Jul05.pdf.

Deutscher, I. 1973. *What We Say/What We Do: Sentiments and Acts.* Glenview, Ill.: Scott, Foreman.

Durant, J., M.W. Bauer, and G. Gaskell, eds. 1998. *Biotechnology in the Public Sphere: A European Sourcebook.* London: Science Museum Press.

European Commission. 1991. Public Transportation and Biotechnology. *Eurobarometer* 35(1).

European Commission. 1993. Energy Policies, Biotechnology, and Genetic Engineering. *Eurobarometer* 39(1).

European Commission. 1997. The Europeans and Modern Biotechnology. *Eurobarometer* 46(1).

European Commission. 2000. The Europeans and Modern Biotechnology. *Eurobarometer* 52(1).

European Commission. 2003. Europeans and Biotechnology in 2002. *Eurobarometer* 58(0).

Fletcher, I. February 10, 1999. Shoppers Demand Clearer Labels on GM Foods. *London Evening Standard,* 7.

Food Standards Agency. September 21, 2001. *Food Concerns Omnibus Survey.* Available at http://www.food.gov.uk/multimedia/webpage/concernsomni.

Food Standards Agency. February 2002. Consumer Attitudes to Food Standards. Available at http://www.food.gov.uk/multimedia/pdfs/consumeratt_uk.pdf

Food Standards Agency. 2004. *Consumer Attitudes to Food Standards Wave 4.* UK Report COI Ref 257960. London: Food Standards Agency.

Gale, F. 2005. China's Food Market Revolution Reaches the Countryside. *Amber Waves* 3(4): 2. Available at www.ers.usda.gov/AmberWaves/September05/pdf/Findings-MTSeptember2005.pdf.

Gaskell, G., M.W. Bauer, J. Durant, and N.C. Allum. 1999. Worlds Apart? The Reception of Genetically Modified Foods in Europe and the U.S. *Science* 285: 384–387.

Giannakas, K., and M. Fulton. 2002. Consumption Effects of Genetic Modification: What If Consumers Are Right? *Agricultural Economics* 27: 97–109.

Greenpeace International. 2003. *The Ingredients Are Hidden, but Companies Can't Hide.* Available at www.greenpeace.org/international/news/the-ingredients-are-hidden-bu.

Hamstra, A.M. 1993. *Consumer Acceptance of Food Biotechnology: The Relation between Product Evaluation and Acceptance.* Research Report 137. Leiden, the Netherlands: SWOKA.

Hamstra, A.M., and C. Smink. 1996. Consumers and Biotechnology in the Netherlands. *British Food Journal* 98(4): 34–38.

Hillers, V.N., and Lowik, M.R.H.. January/February 1998. Incorporation of Consumer Interests in Regulation of Novel Foods Produced with Biotechnology: What Can be Learned from the Netherlands? *Journal of Nutrition Education,* 30(1), 2–7.

Ho, P., and E.B. Vermeer. 2004. Food Safety Concerns and Biotechnology: Consumers' Attitudes to Genetically Modified Products in Urban China. *AgBioForum* 7(4): 158–175.

Hoban, T.J. 1996. Trends in Consumer Attitudes about Biotechnology. *Journal of Food Distribution Research* 27(1): 1–10.

Hoban, T.J. 1997. Consumer Acceptance of Biotechnology: An International Perspective. *Nature Biotechnology* 15: 232–234.

Hoban, T.J. 1998. Trends in Consumer Attitudes about Agricultural Biotechnology. *AgBioForum* 1(1): 3–7.

Hoban, T.J., and J. Burkhardt. 1991. Biotechnology Control of Growth and Product Quality in Meat Production: Implications and Acceptability. In *Proceedings of Determinants of Public Acceptance in Meat and Milk Production: North America Conference*, ed. P. Van der Wal. Wageningen, The Netherlands: Wageningen Agricultural University.

Hoban, T.J., and P.A. Kendall. 1993. *Consumer Attitudes about Food Biotechnology*. Raleigh, N.C.: North Carolina Cooperative Extension Service.

Hornig Priest, S. 2000. US Public Opinion Divided Over Biotechnology? *Nature Biotechnology* 18(9): 939–942.

Hu, W., and K. Chen. 2004. Can Chinese Consumers be Persuaded? The Case of Genetically Modified Vegetable Oil. *AgBioForum* 7(3): 142–132.

Huang, J., S. Rozelle, C. Pray, and Q. Wang. 2002. Plant Biotechnology in China. *Science* 295(5555): 674–676.

Huffman, W.E., M. Rousu, J.F. Shogren, and A. Tegene. 2001. *The Value to Consumers of GM Food Labels in a Market with Asymmetric Information: Evidence from Experimental Auctions*. Paper Presented at the 5th International Consortium on Agricultural Biotechnology Research (ICABR) Meetings, Ravello, Italy.

Huffman, W.E., M. Rousu, J.F. Shogren, and A. Tegene. 2002. *Should the United States Initiate a Mandatory Labeling Policy for Genetically Modified Foods?* Paper Presented at the 6th International Consortium on Agricultural Biotechnology Research (ICABR) Meetings, Ravello, Italy.

Huffman, W.E., M. Rousu, J.F. Shogren, and A. Tegene. 2003. Consumer Willingness to Pay for Genetically Modified Food Labels in a Market with Diverse Information: Evidence from Experimental Auctions. *Journal of Agricultural and Resource Economics* 28(3): 481–502.

Hur, J. March 28, 2001. New GMO Rules May Curb Japan's Appetite for US Corn. *Reuters News*.

IFIC (International Food Information Council). February 2001a. *IFIC Background Report: More U.S. Consumers Expect Benefits: Mixed Feelings, but Not Major Concern Over Labeling*. Available at http://www.ific.org/foodinsight/2001/ma/biotechnbfi201.cfm

IFIC (International Food Information Council). September 2001b. *IFIC Background Report: Most Americans can Articulate Expected Benefits of Food Biotechnology*. Available at http://www.ific.org/foodinsight/2001/so/biotechbenfi501.cfm

IFIC (International Food Information Council). August 2002. *U.S. Consumer Attitudes toward Food Biotechnology*. Available at http://www.ific.org/research/funcfoodsres02.cfm

IFIC (International Food Information Council). April 2003. *IFIC Survey: Americans Acceptance of Food Biotechnology Matches Growers' Increased Adoption of Biotech Crops*. Available at http://www.ific.org/research/upload/IFIC-Survey-Americans-Acceptance-of-Food-Biotechnology-Matches-Growers-Increased-Adoption-of-Biotech-Crops.pdf

IFIC (International Food Information Council). June 2005. *IFIC Survey: Food Biotechnology Not a Top-of-Mind Concern for American Consumers*. Available at http://www.ific.org/research/upload/2005BiotechSurvey.pdf.

Information Office of the State Council. 2000. *White Paper on Population in China*. Available at www.cpirc.org.cn/en/whitepaper.htm.

Kahneman, D., and A. Tversky. 1984. Choices, Values, and Frames. *American Psychologist* 39(4): 341–350.

Kahneman, D., I. Ritov, and D. Schkade. 1999. Economic Preferences or Attitude Expressions? An Analysis of Dollar Responses to Public Issues. *Journal of Risk and Uncertainty* 19(1–3): 203–235.

Kalaitzandonakes, N. 2004. Another Look at Biotech Regulation. *Regulation* 27(1): 44–50.

Kalaitzandonakes, N., R. Maltsbarger, and J. Barnes. 2001. Global Identity Preservation Costs in Agricultural Supply Chains. *Canadian Journal of Agricultural Economics* 49: 605–615.

Keisel, K., D. Buschena, and V. Smith. 2004. Consumer Acceptance and Labeling of Biotech in Food Products: A Study of Fluid Milk Demand. In *Consumer Acceptance of Biotechnology Foods*, ed. R.D. Evenson and V. Santaniello. New York: CABI Publishers.

La Piere, R.T. 1935. Attitudes versus Actions. *Social Forces* 13: 230–237.

Leahy, P.J., and A. Mazur. 1980. The Rise and Fall of Public Opposition in Specific Social Movements. *Social Studies of Science* 10: 259–284.

Li, Q., K.R. Curtis, J.J. McCluskey, and T.I. Wahl. 2002. Consumer Attitudes toward Genetically Modified Foods in Beijing China. *AgBioForum* 5(4): 145–152.

Lusk, J.L., J. Roosen, and J.A. Fox. 2003. Demand of Beef from Cattle Administered Growth Hormones or Fed Genetically Modified Corn: A Comparison of Consumers in France, Germany, the United Kingdom, and the United States. *American Journal of Agricultural Economics* 85(1): 16–29.

Lusk, J.L., L.O. House, C. Valli, S.R. Jaeger, M. Moore, J.L. Morrow, and W.B. Traill. 2004. Effect of Information about Benefits of Biotechnology on Consumer Acceptance of Genetically Modified Foods: Evidence from Experimental Auctions in the United States, England and France. *European Review of Agricultural Economics* 31(2): 179–204.

Marks, L.A., N. Kalaitzandonakes, and S.S. Vickner. 2004. Consumer Purchasing Behavior towards GM Foods in the Netherlands. In *Consumer Acceptance of Biotechnology Foods*, ed. Robert D. Evenson and V. Santaniello. Wallingford, UK: CABI Publishers Ltd. 23–39.

Marris, C., B. Wynne, P. Simmons, and S. Weldon. 2001. *Public Perceptions of Agricultural Biotechnologies in Europe*. Final Report of Perceptions of Agricultural Biotechnologies in Europe (PABE) Research Project 62.

McFadden, D. 1999. Rationality for Economists? *Journal of Risk and Uncertainty* 19(1–3): 73–105.

Menkhaus, D.J., G.W. Borden, G.D. Whipple, E. Hoffman, and R.A. Field. 1992. An Experimental Application of Laboratory Experimental Auctions in Marketing Research. *Journal of Agricultural and Resource Economics* 17(1): 44–55.

Midden, C., A. Hamstra, J. Gutteling, and C. Smink. 1998. The Netherlands. In *Biotechnology in the Public Sphere: A European Sourcebook*, ed. J. Durant, M.W. Bauer, and G. Gaskell. London: Science Museum Press, 103–117.

Moon, W., and S.K. Balasubramanian. 2001. Public Perceptions and Willingness-to-Pay a Premium for Non-GMO Foods in the US and UK. *AgBioForum* 4(3&4): 221–231.

MSNBC. January 2000. *MSNBC Live Vote Results*. Available at www.msnbc.com.

Noussair, C., S. Robin, and B. Ruffieux. 2004. Do Consumers Really Refuse to Buy Genetically Modified Food? *The Economic Journal* 114(492): 102–120.

Nunn, J. 2000. What Lies Behind the GM Label on UK Foods. *AgBioForum* 3(4): 250–254.

Nyborg, K. 2000. Homo Economicus and Homo Politicus: Interpretation and Aggregation of Environmental Values. *Journal of Economic Behavior and Organization* 42: 305–322.

Pennings, J.M.E., B. Wansink, and M.T.G. Meulenberg. 2002. A Note on Modeling Consumer Reactions to a Crisis: The Case of the Mad Cow Disease. *International Journal of Research in Marketing* 19(1): 91–100.

Petts, J., T. Horlick-Jones, and G. Murdock. 2001. *The Social Amplification of Risk: the Media and the Public*. Contract Research Report 329/2001. Norwich, UK: Her Majesty's Stationery Office.

PIFB (Pew Initiative on Food and Biotechnology). March 2001. *Public Sentiment about Genetically Modified Food.* Available at pewagbiotech.org/research/gmfood/survey 3-01.pdf

PIFB (Pew Initiative on Food and Biotechnology). September 15, 2003. *Public Sentiment about Genetically Modified Food: Recent Poll Findings.* Memorandum from the Mellon Group, Inc. and Public Opinion Strategies Inc. to PIFB. Available at pewagbiotech. org/research/2003update/2003summary.pdf.

PIFB (Pew Initiative on Food and Biotechnology). 2004. *Public Sentiment about Genetically Modified Food: Overview of Findings 2004 Focus Groups and Poll.* Available at pewagbiotech.org/research/2004update/overview.pdf.

PIFB (Pew Initiative on Food and Biotechnology). November 7, 2005. *Public Sentiment about Genetically Modified Food: Recent Findings.* Memorandum from the Mellon Group, Inc. and Public Opinion Strategies Inc. to PIFB. Available at pewagbiotech. org/research/2005update/2005summary.pdf.

Piggott, N.E., and T.L. Marsh. 2004. Does Food Safety Information Impact US Meat Demand? *American Journal of Agricultural Economics* 86(1): 154–174.

Prato, T. 1998. *Natural Resource and Environmental Economics.* Ames, IA: Iowa State University Press.

Rousu, M.C., W.E. Huffman, J.F. Shogren, and A. Tegene. 2003. Are United States Consumers Tolerant of Genetically Modified Foods? *Review of Agricultural Economics* 26(1): 19–31.

Rousu, M.C., W.E. Huffman, J.F. Shogren, and A. Tegene. 2004. Estimating the Public Value of Conflicting Information: The Case of Genetically Modified Foods. *Land Economics* 80(1): 125–135.

Rubel, A., and A. Streiffer. 2005. Respecting the Autonomy of European and American Consumers: Defending Positive Labels on GM Foods. *Journal of Agricultural and Environmental Ethics* 18: 75–84.

Runge, C.F., and L.A. Jackson. 2000. Negative Labeling of Genetically Modified Organisms (GMOs): The Experience of rbST. *AgBioForum* 3(1): 58–62.

Sagoff, M. 1988. *The Economy of the Earth.* Cambridge: Cambridge University Press.

Select Committee on Environmental Audit. December 17, 2003. Minutes of Evidence. United Kingdom Parliament, Examination of Witness, Questions 484–499 (testimony of Nick Pidgeon).

Shogren, J.F., J.A. Fox, D.J. Hayes, and J. Roosen. 1999. Observed Choices for Food Safety in Retail, Survey, and Auction Markets. *American Journal of Agricultural Economics* 81(5): 1192–1199.

Smith, M.E., E.O. van Ravenswaay, and S.R. Thompson. 1988. Sales Loss Determination in Food Contamination Incidents: An Application to Milk Bans in Hawaii. *American Journal of Agricultural Economics* 70(3): 513–520.

Smith, V.L. 1991. Rational Choice: The Contrast between Economics and Psychology. *Journal of Political Economy* 99(4): 877–897.

Streiffer, R., and A. Rubel. 2003. Comment: Choice versus Autonomy in the GM Food Labeling Debate. *AgBioForum* 6(3): 141–142.

Swartz, D.G., and I.E. Strand, Jr. 1981. Avoidance Costs Associated with Imperfect Information: The Case of Kepone. *Land Economics* 57(2): 139–150.

Tolley, G.S., and A. Randall. 1983. *Establishing and Valuing the Effects of Improved Visibility in the Eastern United States.* Washington, D.C.: U.S. Environmental Protection Agency.

van Ravenswaay, E.O., and J.P. Hoehn. 1991. The Impact of Health Risk Information on Food Demand: A Case Study of Alar in Apples. In *Economics of Food Safety*, ed. J.A. Caswell. New York: Elsevier Science, 155–174.

van Wechel, T., C.J. Wacheheim, E. Schuck, and D.K. Lambert. 2003. *Consumer Valuation of Genetically Modified Foods and the Effect of Information Bias.* Agribusiness and Applied Economics Report No. 513. Fargo: Department of Agribusiness and Applied Economics, North Dakota State University.

Vidal, J. June 4, 1998. Gene Genie. *The Guardian*, 14.

Walsh, J. January 11, 1999. Brave New Farm. *Time Magazine,* 86–87.

Wessels, C.R., C.J. Miller, and P.M. Brooks. 1995. Toxic Algae Contamination and Demand for Shellfish: A Case Study of Demand for Mussels in Montreal. *Marine Resource Economics* 10: 143–159.

# 8

# Frankenfood Free

## Consumer Sovereignty, Federal Regulation, and Industry Control in Marketing and Choosing Food in the United States

*Thomas O. McGarity*

One of the most contentious of the many public policy debates over foods resulting from modern biotechnology has been the debate over whether genetically modified (GM) foods should be labeled so that consumers can easily ascertain whether the food that they are consuming has been genetically modified or contains GM constituents. Governments in Europe have resolved the debate largely in favor of labeling, while the federal government in the United States has declined to require labeling (Grossman 2005: 45). Congress has considered bills requiring labeling on foods containing GM materials on many occasions but has thus far failed to enact any of them (Galant 2005: 153). Voters in Oregon considered, and rejected, an initiative requiring such labeling in 2002 (Mayer and Cole 2002). A court of appeals decision involving milk produced through the application of a genetically engineered bovine growth hormone even casts some doubt upon the constitutionality of state and perhaps even the federal labeling requirements (*International Dairy Foods v. Amestoy* 1996). Yet the debate rages on.

This chapter provides an overview of the policy debate over the legal and regulatory issues involved in labeling GM foods. It then describes how the courts and the Food and Drug Administration (FDA), the U.S. regulatory agency with the most direct role in determining what manufacturers can say to consumers about health-related aspects of food, have resolved (or failed to resolve) those legal issues with an eye toward how the current legal regime advances or detracts from several important policy considerations that have arisen during the debates.

### THE POLICY DEBATE

One of the continuing conundrums of a market economy is the question of who provides how much and what kind of information to consumers. In a strictly laissez faire regime, caveat emptor is the rule, and advertising is the primary informational vehicle. It is up to the consumer to acquire at the consumer's expense such additional information as the consumer deems necessary to evaluate the quality and safety of the product. If the product is unsafe, if the product is not what its manufacturer holds it out to be, or if the product or the manufacturing process used to produce it is environmentally detrimental, the consumer may elect not to purchase it, but the burden is on the consumer to ascertain (perhaps with the help of a private entity such as the Consumers Union) its relevant features.

In a mass market economy, accurate information about consumer products is critical to ensuring public trust in the marketplace, and modern democracies have generally abandoned the strict laissez faire model for a regulatory regime in which government agencies play a role in protecting consumers from fraud and misleading advertising (Beaudoin 1999: 239–340). Informational requirements can, of course, be burdensome and, in the extreme, may discourage technological advance. Product label requirements can also be both economically and administratively inefficient if the value of the information to consumers does not exceed the cost of the additional label and the opportunity cost of the information that otherwise might be conveyed in the limited space available. Thus, there may remain a domain of information that is indisputably relevant to consumer choice that is nevertheless appropriately left to the marketplace. Reasonable people may disagree on the extent to which government should force manufacturers to provide all consumers with information that some consumers might deem relevant to their purchase decisions, but is not needed to prevent deception and is not directly related to latent dangers. There is, however, little disagreement over core labeling requirements such as those designed to protect persons with food allergies and other special needs, and the current legal regime implements those core requirements fairly effectively.

## Introduction to the GM Labeling Debate

Amid the multifaceted policy debates over labeling GM foods, two broad legal issues predominate: (1) whether the law should require companies marketing foods to provide information concerning the presence of GM materials in those foods on the labels, and (2) whether the information that companies voluntarily include on food labels concerning the presence or absence of GM materials is misleading. Since the resolution of both of these questions invokes sensitive and powerful concerns about the power of the state to regulate communication between marketers and consumers, an overarching legal issue is the constitutionality of any legal requirements imposed by Congress, a state legislature, or a regulatory agency.

Few would disagree with the proposition that accurate labeling is essential to informed consumer choices about the quality, safety, and other important aspects of food. To the extent that the unimpeded market will provide accurate information concerning the presence and absence of GM materials in food, it should be allowed to do so in terminology that is comprehensible to ordinary consumers, even if the terminology is inaccurate or imprecise as a purely scientific matter. If the market will not provide that information voluntarily, then the government should insist that manufacturers do so through appropriate legal requirements.

Biotechnology proponents, however, believe that the regulatory agencies in the United States have wisely refrained from requiring labeling for GM foods, because those foods have been tested with greater thoroughness than most foods in the U.S. diet. Since extensive testing has, in their view, revealed no "scientific" basis for distinguishing between GM and non-GM foods on safety or environmental grounds, labeling requirements are an attempt to "fix a problem that does not exist." Only when genetic modification of traditional foods results in significant changes in their nutritional or safety status should separate labeling of GM foods even be considered (Michael Phillips, quoted in "Industry Opposes Biotech Labeling" 2000). On the other hand, the government should require manufacturers that claim that their foods are free of GM materials (or that they have undertaken to render them free of GM materials) to do so in language that is technically accurate and does not in any way imply that food not so labeled is in any way inferior.

## Food Safety

Genetically engineered food can be unsafe, either because it has been engineered to contain toxic substances or because a gene coding for a toxic substance has been inadvertently inserted into the plant (McGarity 2002: 417). The magnitude of the direct risks posed by a GM food item is a function of the likelihood that the inserted gene will code for a substance that is toxic to some or all human beings, the nature and potency of that substance's toxicity, and the extent of dietary exposure to that substance (National Academy of Sciences 2000: 55). Since GM pest-resistant plants are designed to be toxic to the target organisms, their potential to affect human health adversely is of obvious concern. In addition, modern genetic engineering techniques have the potential to change internal plant metabolic processes in several ways that render them riskier for consumers (McGarity 2002: 417). The genetic engineers also might inadvertently produce a plant in which the levels or bioavailability of important nutrients are altered in significant ways that could be harmful to human health (FDA 1992: 22987). Finally, it is possible that an inserted gene will code for a substance that is allergenic to some subpopulation of sensitive individuals (McGarity 2002: 417). Although sensitive persons can usually minimize the risk of allergenic responses by avoiding particular foods, they may be caught unawares if GM food manufacturers transfer the genes that code for allergenic proteins from one food plant to another (Royal Society of Canada 2001: 53).

A cautious manufacturer will attempt to assess the health risks posed by any new proteins, carbohydrates, fats, or oils in GM foods before putting them on the market. Yet, in many cases, it is not clear how manufacturers should go about testing for the presence of unanticipated toxins. Since scientists do not necessarily know which genes code for allergenic proteins, they will not always know whether they have accomplished such a transfer. For example, scientists working for Pioneer Hi-Bred transferred into soybeans a gene from Brazil nuts in the hope of creating a soybean with a higher nutritional value. The experiment produced a soybean that was allergenic to people who were allergic to Brazil nuts. Fortunately, the company then decided voluntarily to stop most of its work on the product (National Academy of Sciences 2000: 67). The unfortunate reality is that substantial uncertainties permeate the existing state of knowledge regarding the risks and benefits of GM foods. The easiest way for a consumer to avoid such risks is to avoid eating foods containing GM materials. And that simple strategy is unavailable if foods containing GM materials are not labeled as such.

The biotechnology industry and the current leadership of the relevant U.S. regulatory agencies have gone to great lengths to persuade the public that modern GM foods pose minimal direct risks to human health. Biotechnology proponents insist that any direct toxicity risks to human health are trivial and unworthy of serious consideration. Biotechnology companies have no incentive to design GM plants to be toxic and every incentive to avoid that outcome. Genetically modified plants go through many iterations of testing and analysis during their development, and at each stage of development, manufacturers take great pains to ensure that undesirable toxins have not been inadvertently generated (McGarity 2002: 417).

Critics point out that these assurances are not based upon extensive testing of GM foods in laboratory animals as in the case of chemical additives and pesticides, but instead depend heavily upon assumptions derived from knowledge about the risks posed by non-GM plants. Indeed, the relevant regulatory agencies in the United States have not required extensive testing of GM foods but have instead relied upon the doctrine of "substantial equivalence" to bypass the need for extensive testing. Put simply, they have concluded that if a GM food is "substantially equivalent" to a non-GM food that has a history of safe use, the GM food should not be regulated any more stringently simply because it is the product of modern biotechnologies (Lappe and Bailey 1998: 29–32; Engel et al. 1995: 7).

Critics believe that modern gene splicing techniques are sui generis and therefore reject the major premise underlying the "substantial equivalence" doctrine. Noting that modern gene splicing tools can accomplish in a single generation changes that would take millennia or could never occur in nature, critics worry about the exercise of scientific judgment that goes into the determination of both "equivalence" and "substantiality." The legitimacy of substantial equivalence depends heavily on who is determining equivalence and the criteria that they employ. The critics do not trust either the manufacturers, who make that determination in the first instance, or overworked and sometimes overly solicitous bureaucrats, who make the final determination, to employ the substantial equivalence doctrine in a way that adequately protects the health of consumers. This level of mistrust is not necessarily unreasonable. As both the assigned promoter of U.S. agricultural products and the regulator of genetically engineered plants (at least prior to general release into the marketplace), the U.S. Department of Agriculture (USDA) labors under an institutional conflict of interest (Engel et al. 1995: 3–4; Marden 2003: 773–774).

Many consumers would rather have the option of protecting themselves by electing not to purchase foods containing GM materials. They may simply want to await further testing of or broader experience with GM foods before consuming them. Even if GM foods pose only tiny health risks, many consumers see no reason why they should be forced to assume those risks without their consent. Consumers with severe allergies or special dietary limitations have a special interest in avoiding foods that are not "tried and true" on the theory that it is better to be safe than sorry. Avoiding foods containing GM materials, however, requires labeling sufficient to inform consumers of their presence (Kirby 2001: 358–359).

Opponents of labeling flatly dispute the contention that foods containing GM materials are any less safe than other foods. Recognizing the possibility that powerful genetic engineering techniques are certainly capable of producing, either intentionally or inadvertently, foods that are more toxic or allergenic than the foods that consumers normally eat, they argue that it is the job of the relevant regulatory agencies to ensure that such foods do not get out on the market, and they see no evidence that the FDA is failing in that regard. If any particular GM food is in fact likely to be more toxic, the relevant regulation agency should address its potential toxicity directly by banning or otherwise regulating the food. Labeling GM food products is the wrong way to go about addressing that toxicity. If allergenicity is a concern, then "the crucial information is the nature of the added allergen, not the means by which it was incorporated in the product" (Beales 2000: 108). The fact that food contains GM materials is not, in their view, a useful surrogate for the presence or absence of an introduced allergen (Leggio 2001; Beales 2000).

### Food Purity

Consumers may have concerns about the "purity" of the food that they eat that go beyond direct safety concerns. People may want to eat "organic" food not so much because it is safer but because they believe that it is purer and generally more wholesome than foods that have been treated with pesticides, processed using unnatural techniques, or otherwise impure. The huge outcry that attended the proposal by the USDA to include GM foods within the category of organic food indicates the intensity of consumer preference for foods that are not in this sense "unnatural."

Labeling opponents reject food purity as a relevant consideration insofar as it purports to differ at all from food safety. "Natural" foods are not necessarily pure, if by "pure" we mean unlikely to cause disease. And if "pure" is to be given any other meaning, then it rapidly begins to take on religious connotations, and arguments advancing food "purity" should be acknowledged as essentially spiritual in nature and addressed in a debate over religious considerations. On the other hand, if "pure" and "organic" are coterminous, then food purity considerations as applied to GM foods can easily be folded into the definition of "organic" by excluding GM foods from the category of "organic" food.

### Environmental Protection

Some consumers may want to avoid GM foods out of a personal desire to discourage the trend toward environmentally damaging monocultures and to reduce the risk of destroying indigenous native species of those plants through "cross-fertilization" with GM crops. For example, many of the strong proponents of labeling milk from cows that have been treated with bovine growth hormone recombinant bovine somatotropin (rBST) were not so concerned with the direct health effects of the genetically engineered hormone as they were concerned about the indirect health effects of antibiotic resistance resulting from the tendency of farmers to provide heavy doses of antibiotics to rBST-treated cows to prevent mastitis (Burk 1997). Voting with one's pocketbook is an entirely legitimate and often quite effective way for a consumer to take direct action to protect the environment (Kysar 2001). One cannot decline to purchase GM foods, however, if they are not labeled as such.

Like their response to the food safety rationale, labeling opponents argue that Congress has enacted laws and created agencies for the purpose of protecting the environment from the adverse effects of private sector products and activities. If people want the government to do more, they should act as citizens to secure the enactment of more stringent environmental protection laws and regulations and not as consumers to boycott products that they believe, perhaps erroneously, will harm the environment. Enacting labeling laws not only adds to the expense of GM foods for those who do not believe that they will cause environmental harm (or do not care if they do), but also intrudes on the rights of manufacturers to communicate with the public without unnecessary governmental interference.

## Precaution

Many modern consumers who are not at all unaware of the true nature of health and environmental risks and are by no means uninformed are, to the contrary, very much aware of instances in the past where the proponents of technologies were, in retrospect, far too optimistic about their benefits and far too anxious to downplay their hazards. A consumer who knows nothing about genetic engineering may know a lot about how "miracle" drugs have caused catastrophic injuries, how nuclear power created a legacy of radioactive waste, and how the kudzu plant that was imported as an erosion control tool has taken over the rural South. One does not have to have an extensive knowledge of the risks and benefits of a technology to have a well-founded skepticism of its industrial and governmental proponents and a rational apprehension about what they do not know (or are not telling) about their products.

Although this is not the place for an extended discussion of the pros and cons of the Precautionary Principle, suffice it to say that opponents of labeling are generally not big fans of the Precautionary Principle. The Precautionary Principle, in the view of its critics, has no defined limits. It represents little more than a vague injunction to "err on the side of safety" without providing any guidance as to how much uncertainty is tolerable and how many valuable technologies society should forgo merely to avoid the perils of the unknown (Sunstein 2003). Moreover, uncertainty about toxicity is not unique to GM foods. Conventional plant breeding techniques also present the very real possibility that something will go wrong and the resulting food plant will be toxic or allergenic (Beales 2000: 110).

## Religion and General Moral Considerations

Many consumers may have moral or religious reasons for avoiding GM foods that contain genes from particular species of host organisms. For example, many religions have important things to say about what foods practitioners may eat and when they eat them. Science did not dictate the Jewish religious practice of refraining from eating pork or the Catholic religious practice of refraining from eating red meat on Friday. Vegan Buddhists may want to know whether vegetables have been genetically engineered to contain genes coding for animal proteins. Moral concerns less precise than a religious ban on eating pork may legitimately motivate a consumer to want to know whether a particular piece of food has been genetically modified. Consumers may, for example, be "morally opposed to GM foods, believing that they are 'incompatible with the integrity of nature'" (Kirby 2001: 357).

Labeling opponents respond that the food needs of the practitioners of a particular religion might justify tolerance on the part of the state for voluntary labeling in accordance with religious practices, but they cannot justify an intrusion of the state into the marketing practices of companies that prefer not to cater to particular religious practices. To the argument that vegans might object to eating plants containing animal genes, for example, opponents of labeling respond that "the resulting genetic combination is a plant, behaves as a plant, and reproduces as a plant" and is no more animallike than "a plant fertilized with manure" (Beales 2000: 110). The protein that the transported DNA calls for may well be present in other plant species (Beales 2000: 110). It would therefore be irrational for a vegan to object, for religious reasons or otherwise, to eating plants containing animal genes.

## Consumer Right to Know?

Some labeling advocates argue that labeling GM foods is a necessary manifestation of a broader "consumer right to know." For example, the consumer group Consumers Union maintains that "[c]onsumers have a fundamental right to know what they eat, and federal officials should require that all foods containing genetically engineered ingredients be labeled as such" ("Seeds of Change" 1999: 46). Opponents warn that the "right to know" knows no limits. In theory, such a right "could be invoked to justify labeling about any detail of the production process" (Beales 2000: 109). Rather than involving the state in defining and enforcing a vague and indeterminate right to know, labeling skeptics argue, the government should let the marketplace decide the extent to which information about food is conveyed to consumers. To the extent that consumers want to know information about the content of food or the processes used to make it, the market can be expected to fulfill that desire as it has done for centuries with kosher foods and, more recently, organic and "fair trade" foods. Consumers who are willing to pay for the knowledge and for any extra procedures that must be undertaken to ensure that the food is in fact free of GM materials will purchase the labeled foods at correspondingly higher prices, but consumers who do not care can purchase lower cost food containing GM materials (Barrett 1999; Degnan 2000: 310).

## Informational Accuracy

One serious disadvantage to allowing the market to decide what information goes on food labels is the very real possibility that manufacturers will put misleading information on the labels. Food and drug labeling has a long and sordid history of manipulation by unscrupulous companies to overcome consumer resistance and to gain competitive advantage. Even in a world in which consumers can read *Consumer Reports* and freely search the Internet, manufacturers have an enormous informational advantage over consumers, and they can be expected to use it to their economic advantage.

Labeling proponents believe that many consumers think they are eating foods that have not been genetically modified when in fact the foods have been genetically engineered. It is therefore inaccurate to market food containing GM materials as if it were not genetically modified. As GM food crops begin to predominate American agriculture, this argument will lose its persuasiveness. If most people are aware of the fact that there is a high probability that certain kinds of foods have been genetically modified, then the expectations will shift and the presumption will be that food has been genetically engineered unless the label says otherwise (Kirby 2001: 356).

When manufacturers attempt voluntarily to inform consumers that their products do not contain GM modified materials, informational accuracy remains a serious issue. Agricultural biotechnology advocates argue that the representation on a label that the contents are free of GM materials has a great potential to "mislead" or "confuse" consumers. They note that all modern crops are genetically modified in the sense that they have resulted from traditional plant breeding technologies that over several generations change the genetic structure of the seeds that are sold for planting by farmers. Yet no one has suggested that hybrid corn and similar crops should be labeled as such. Furthermore, a labeling requirement would send an implicit message to consumers that they should be worried about GM foods because they are less healthy or otherwise inferior. The words "Frankenfood Free" on a label, for example, would certainly suggest that food not so labeled contains something monstrous, when in fact most scientists believe that food containing GM materials is no more hazardous than unmodified foods (Howle and Sim 1998: 25; Beales 2000: 107; Miskiel 2001: 238).

Some opponents of GM labeling would go farther, however, to take the position that the potential for misleading is so great that the government should not allow manufacturers to say anything about the presence or absence of GM materials beyond precisely what the government requires to protect health and prevent fraud. In their view, consumers are badly informed by the media about the hazards of GM foods or are otherwise ignorant of the benefits and risks and are therefore not qualified to decide whether or not GM foods are good for them. Since government agencies such as the FDA and USDA are qualified to make such determinations, the government should decide for them that they do not need to know about the presence of GM materials in their food. Like the so-called "therapeutic privilege" that excuses the failure of a doctor to secure the informed consent of a patient who might use the information detrimentally, the government should keep ill-informed consumers ignorant for their own good (Goldman 2000: 721–722).

Consumer advocates reject the argument that accurate labeling might "mislead" consumers as paternalistic and wholly inconsistent with the notion of "consumer sovereignty" that lies at the heart of a market-based economy. Many consumers are offended by the implicit (and sometimes explicit) assertion that food processors know what is best for their consumers. If the industry and the agencies want consumers to trust agency safety decisions reached on the basis of industry-provided health and safety data, then the industry and the agencies must be willing to trust consumers to make informed decisions on the basis of reasonable labeling requirements. Companies are certainly capable of using advertising to make their case that GM foods are as safe as non-GM food, and the relevant federal agencies can conduct appropriate consumer research to determine whether consumers are being mislead by mandatory labels (Groth 1999; Jacobson 1999).

### Information Overload

As food manufacturers are required to put more and more information on food labels, critics of GM labeling requirements fear that consumers suffering from "information overload" will give up and ignore all of that information (Miskiel 2001: 248). Too much information of a trivial nature will "crowd out more important information and blunt the intended and desired impact of such information" (Degnan 2000: 306). In promulgating regulations implementing the Nutrition Labeling and Education Act of 1990, the FDA noted that "not all information related to maintaining healthy dietary practices can be included on the food label," because "the large amount of information would interfere with consumers' zabilities to use the information of the greatest public health significance" (FDA 1993a: 2107). Given the salutary track record of GM foods, GM labeling opponents maintain that information about the presence or absence of GM materials is clearly not material and should therefore not be included on food labels.

### Segregation

A voluntary label implying that a food is "GMO free" or contains "No GMO" provokes its own unique debate over whether any such representation can be made with sufficient assurance to avoid misleading consumers. Labeling opponents argue that a label implying that a food has no GM material may well be factually inaccurate and therefore misleading to consumers. Genetically modified plantings for some crops, such as soybeans, have become so ubiquitous in the United States and the tools for segregation so weak that it may be impossible for a company to guarantee that the materials that it uses in food do not contain any GM material (Endres 2005: 132–133). The industry points to practical difficulties in keeping GM grain separate as it moves from planting to field to elevator to processor to groceries (Goldman 2000: 721). Labeling opponents demand that any explicit claims that a food is GM-free and any implicit claims that GM-free foods are superior to GM foods be substantiated with scientific evidence to protect consumers from fraud (Frank 1999; Keith 1999).

Labeling advocates, however, point out that farmers "have long practiced variations of modern segregation and identity preservation" to reduce the likelihood of adventitious mixing of undesirable and desirable food constituents (Endres 2005:135). Indeed, several well-known companies in the United States have developed procedures for segregating non-GM foods for the very practical reason that they want to market their products in Europe, which subjects GM food to a labeling requirement. Even if it is impossible to ensure that a piece of food does not have a single molecule of modified DNA in it, consumers understand that the label "GMO free" represents an assurance that the manufacturer has employed the best available techniques and technologies to ensure that the food item does not contain any GM materials. If that is still misleading to consumers, the solution is not to abandon labeling but to allow for "adventitious" amounts of GM material of less than a prescribed percentage, and the European Union has done precisely that (Endres 2005: 135, 144–48; Kirby 2001: 363).

### Practical Considerations

Opponents of GM labeling also raise other practical considerations as a barrier to mandatory labeling. One clear practical limitation on labeling requirements is the amount of space realistically available on the label for a food item. Since it is impossible for manufacturers to put on a label of manageable size every item of information that every consumer might deem relevant, the limited space on food labels should be used to convey important nutritional and safety information, not trivial or inconsequential facts (Bohrer 1994: 659; Beales 2000: 114). As with information overload, however, the issue is whether the GM status of the contents of food is of sufficient materiality to warrant inclusion in that limited space. Citing strong consumer interest in that information, GM labeling advocates argue that it is.

The most important practical constraint on any entity subject to a regulatory requirement is the cost of compliance. Unless the requirement is the highly unrealistic demand that the manufacturer determine definitively that the food either does or does not contain GM materials (in which case testing or segregation would be required as well), the cost of labeling is the rather trivial one of changing the label to read that it "may contain" GM materials. Even that modest expense can be substantially reduced if the government phases in the labeling program to provide sufficient lead time to design the information into new labels.

### Efficiency

If we view efficiency considerations broadly to connote utilitarian concerns for avoiding waste, then a policy (in this case labeling GM foods) should be implemented if the beneficiaries of the policy gain more than those harmed by the policy lose. Opponents of mandatory labeling argue that a legal requirement that foods containing GM materials be labeled would fail this test. Citing studies conducted by North Carolina State University scientist Thomas Hoban, opponents argue that the vast majority of consumers are simply uninterested in whether food items contain material from GM plants (Hoban 1998: 70). Yet they will have to pay their share of the extra cost of labeling. The added benefit of labeling devolves to the few consumers who want to know whether it contains GM material. Under a voluntary system, by contrast, those consumers who are willing to pay for the segregations effort that goes into ensuring that food does not contain GM materials will pay extra for labeled foods, and those who do not value the information that highly will purchase food that may contain GM material (Beales 2000: 113).

Consumer advocates are generally wary of the efficiency criterion, and they typically oppose quantitative cost–benefit analysis as the decision criterion for consumer safety regulation because it tends in practice to favor business interests over consumer interests. However, to the extent that efficiency considerations are on the table, they argue that uniform federal labeling requirements would be far more efficient than sporadic and inconsistent market-driven efforts by some companies to label products as GM free. (McGarity et al. 2004; Groth 1999). They interpret the Hoban studies differently, and they cite different polls to suggest that consumers generally desire accurate GM labeling. For example, in a poll undertaken by the biotechnology firm Novartis, 93% of the respondents agreed that GM foods should be labeled, and 73% strongly agreed with that proposition. Similarly, an ABC News poll determined that 93% of the respondents thought the government should require GM labeling (Zeichner 2004: 485).

Biotech proponents dismiss such polls as reflecting a general consumer ignorance of the fact that people already eat many foods that have been modified through traditional breeding techniques. Moreover, they point out that when put to the test in the marketplace, rather than sometimes misleading polls, consumers in much smaller numbers elect to pay more for foods that are free of GM materials. Pointing out that very few labeled GM foods are available for purchase in the European Union even at lower prices, labeling opponents suspect that mandatory labeling requirements are really a "Trojan horse" put into place by anti-GM advocates to facilitate campaigns against all GM foods. Fearing such campaigns, food processors and retailers avoid GM foods altogether, rather than risk the ire of the activists and the resulting loss in sales (Carter and Gruere 2003: 70). Yet, whatever the merits of this view in the context of the European Union, it probably overestimates the power of somewhat marginal activist groups in the United States to cow such powerful economic actors as General Foods and Wal-Mart.

## THE CURRENT REGULATORY REGIME

Three agencies of the federal government have played primary roles in regulating the development and marketing of foods containing GM plants. The USDA regulates the beginning phases of the development of GM crops up to the point that they become commercially viable, at which point they leave the USDA's regulatory domain. The U.S. Environmental Protection Agency (EPA) regulates the marketing and labeling of pesticides, and it has determined that plants that have been genetically modified to contain genes coding for pest resistance are pesticides subject to EPA regulation (McGarity 1987). Although its regulatory program for pesticides addresses the labeling that accompanies packages of GM seeds that farmers purchase, the EPA has not attempted to expand its authority to reach the marketing of the resulting crops to consumers. Only the FDA has directly addressed consumer labeling of GM foods, and that program has evolved over the last 15 years.

## Mandatory Labeling

Under section 403(a)(1) of the Food, Drug and Cosmetics Act of 1938 (FDCA), a food is misbranded if its labeling is "false or misleading in any particular" or if its labeling does not prominently feature "any word, statement, or other information" that the FDA lawfully requires with "such conspicuousness (compared with other words, statements, designs, or devices, in the labeling) and in such terms as to render it likely to be read and understood by the ordinary individual under customary conditions of purchase and use" (§ 343(a)(1), (f)). A food is also misbranded if its label does not bear "(1) the common or usual name of the food, if any there be, and (2) in case it is fabricated from two or more ingredients, the common or usual name of each such ingredient" (§ 343(i)). Under section 201(n) of the FDCA, labeling is misleading if it fails to reveal all facts that are "material in light of…representations or material with respect to consequences which may result from the use of the article to which the labeling…relates under the conditions of use prescribed in the labeling…or under such conditions of use as are customary or usual" (§ 321(n)). Thus, under section 201(n) "both the presence and the absence of information are relevant to whether labeling is misleading" (FDA 2001).

The statute appears to grant the FDA ample authority to require manufacturers and importers to label all food containing GM materials. The agency could, for example, conclude that the fact that a food contains materials derived from GM plants is "material" in light of implicit representations that it is what it appears to be (i.e., food derived from nonengineered plants). In addition, sufficient uncertainties surround the health and ecological "consequences" of GM plants that the agency could reasonably conclude that the fact that a food derives from GM plants is material in light of those consequences. The agency has exercised its powers narrowly, however, to require labeling primarily in cases where important nutritional aspects of GM plants might differ from those of the host plants.

In a major policy statement issued in May 1992, the FDA specifically rejected the suggestion that it require labels for all foods containing GM constituents (FDA 1992: 22986). The agency took the position that modern genetic engineering techniques were merely "extensions at the molecular level of traditional methods," and it observed that the agency had not previously considered the methods used in the development of a new plant variety "to be material information within the meaning of section 201(n)." Since the agency was "not aware of any information showing that foods derived by these new methods differ from other foods in any meaningful or uniform way, or that, as a class, foods developed by the new techniques present any different or greater safety concern than foods developed by traditional plant breeding," it did "not believe that the method of development of a new plant variety is normally material information" that must be disclosed in labeling for the food (FDA 1992: 22991).

The FDA's 1992 legal position is open to serious criticism, both as a matter of statutory interpretation and as a matter of consistency with past agency interpretations of the same language. Public comments on the policy statement noted that in connection with irradiated foods the agency had concluded that "[w]hether information is material under section 201(n)...depends not on the abstract worth of the information but on whether consumers view such information as important and whether the omission of label information may mislead a consumer" (FDA 1986). The agency later explained that in the case of irradiated foods, it concluded that radiation could in some cases cause changes in the organoleptic properties of the finished food (taste, color, feel, etc.) that, absent labeling, might mislead consumers into assuming that such foods were unprocessed (FDA 1993b: 25838). The agency did not explain how this distinguished GM foods, some of which also experience organoleptic changes as a result of the genetic engineering technology. In the case of irradiated foods, the FDA clearly extended the labeling requirement to a class of foods defined by a processing technology and not by the nature of the product.

In addition to its strained "scientific" reasoning, the agency's explanation did little to clarify why changes brought about by genetic engineering techniques that clearly affect the characteristics of the modified foods would not be viewed as important by consumers. As discussed above, labeling proponents believe that an indication on the label that a food contains GM materials is necessary to prevent the public from being mislead (Beaudoin 1999: 272). Indeed, genetic changes may well render the "common name" of a GM food not properly descriptive. A GM potato containing genes coding for proteins that are not otherwise produced by potatoes (e.g., a protein derived from salmon) is intended to be different and is therefore arguably not really a potato to the typical consumer. If the GM food is different from an existing non-GM food to the average consumer, then the statute appears to require proper labeling.

The FDA's sister agency, the USDA, proved quite willing to protect consumer expectations in the rule making that it undertook to promulgate standards under the Organic Foods Production Act of 1990, for organically grown foods in 2001. The USDA's original proposal, which would have allowed foods containing GM materials to be marketed as "organic," generated more than 275,000 public comments, most of which were strongly negative. Heavily featured in the comments was the argument that allowing GM materials to be marketed as "organic" would violate consumer expectations.The USDA accepted the criticism, and the final rules define genetic engineering as an "excluded method" ( USDA, December 21, 2000. ). The department made no attempt to hide the fact that it had changed its approach in order to meet consumer expectations about the "naturalness" of organic food. The American Crop Protection Association, an agribusiness industry trade association, expressly accepted this result, but "only with the clear understanding that the organic designation is in no way an indication of safety or quality but is rather a marketing standard" (Zeichner 2004: 484).

The FDA's 1992 policy statement did acknowledge a limited roll for labeling in the context of GM foods. The agency read section 403 to require manufactures to inform consumers "by appropriate labeling, if a food derived from a new plant variety differs from its traditional counterpart such that the common or usual name no longer applies to the new food, or if a safety or usage issue exists to which consumers must be alerted" (FDA 1992: 22991). In particular, the agency might require labeling to put sensitive subpopulations on notice of the possibility that a transfer from an allergenic donor plant to a previously nonallergenic food plant might have transferred a gene coding for an allergenic protein (FDA 1992: 22987). For example, labeling might be required for a tomato that has been genetically engineered to produce a peanut protein absent "sufficient information to demonstrate that the introduced protein could not cause an allergic reaction in a susceptible population" (FDA 1992: 22991).

On the other hand, the 1992 policy statement did not propose to do anything about the possibility that a genetically engineered crop might contain a brand new protein not produced in significant amounts in either the source or recipient plant that might be allergenic to a sensitive subpopulation. Since the agency was "unaware of any practical method to predict or assess the potential for new proteins in food to induce allergenicity," it was apparently willing to assume that such changes could never occur (FDA 1992: 22987). This posture deprives consumers, who are likewise unaware of practical methods to assess the allergenic potential of such foods, of the option of playing it safe by avoiding such foods.

The agency also recognized that in the past it had required labeling to include the origin of an added ingredient when that information would be material for "religious or cultural reasons," but it cautioned that it had also rejected requests to include the origin of ingredients despite arguments that religious and cultural considerations made the origin material for these reasons (FDA 1993b: 25838). The agency explained that the cases in which it had required labeling for "religious or cultural" reasons involved "the addition to food of food ingredients directly derived from animal or microbial sources." GM foods were different because "the presence in a plant chromosome of deoxyribonucleic acid (DNA) that was originally derived from an animal or microorganism...is now an inherent constituent of a plant." The agency explained that "[o]nce the copies are transferred to the plant, they become an integral part of its genetic information, just like thousands of other genes that are present in the plant chromosome." As a scientific matter, the agency concluded that such transfers did not "confer 'animal-like' characteristics to the plant" (FDA 1993b: 25838).

This rather pained "scientific" explanation seems wholly irrelevant to the question whether for "religious or cultural" reasons the fact that a plant contains a deliberately added gene from a bacterium or an animal is "material" within the meaning of the statute. Science cannot dictate what religion or culture deems to be "material" any more than religion or culture can dictate scientific principles. Once the agency accepted the proposition that religious or cultural considerations could affect the materiality of the addition of an ingredient to food, it could not logically close the door to religious or cultural considerations that it deemed scientifically unwarranted. To an individual who for religious reasons objects to eating human flesh, the fact that a food plant has been genetically engineered to include a human gene coding for a protein found in human skin might be material, despite the FDA's scientific assurance that the plant did not thereby become human. Similarly, a religious order might well determine that eating "planimals" is just as objectionable as eating animals.

The FDA has continued to adhere to the "substantial equivalence" doctrine that permeated the 1992 guidelines. The practical effect of that doctrine is to allow manufacturers of GM foods to avoid any safety testing beyond that necessary to demonstrate substantial equivalence. The legal rationale for this position is that the process used to create the new variety of food, be it genetic engineering or traditional plant breeding techniques, is simply not "material" if the products are substantially equivalent, and the FDA therefore lacks the authority to require labeling (Degnan 2000: 309). The substantial equivalence showing itself is surprisingly subjective, relying essentially upon the "scientific judgment" of those making the determination rather than on any prescribed methodology or series of tests. A federal district court in *Alliance for Bio-Integrity v. Shalala* (2000) upheld the agency's position on labeling GM foods.

## Voluntary Labeling

The FDA unquestionably has the authority to prevent food manufacturers from making misleading claims on the labels of their products ( Food, Drug and Cosmetics Act of 1938 §§ 321(n), 343(a)). Insofar as voluntary labels contain claims about the presence or absence of GM materials that are misleading, the FDA can seize the offending products. In the context of genetic engineering, the agency's primary concern has been for protecting consumers from misleading claims that GM foods are unsafe or otherwise inferior to non-GM foods (Degnan 2000: 308). The FDA's approach to labels on milk containing rBST, a hormone that is given to cows to enhance milk production, is illustrative.

Because it was a food additive, the FDA approved rBST for use in cows in 1993, but it declined to require that labels on milk coming from cows treated with rBST make the consumer aware of that fact (Burk 1997). The FDA later provided guidance on voluntary labeling for rBST. The FDA agreed with the manufacturer of rBST that claims on a milk label that the contents were "BST-free" would be misleading because all milk contains BST. Even a statement that the milk came "from cows not treated with rBST" would be misleading because it might "imply that milk from untreated cows is safer or of higher quality than milk from treated cows" (FDA 1994: 6280). The agency found no evidence to distinguish rBST from BST on health or purity grounds. The FDA's guidance gave both sides something by allowing labels to indicate that the cows producing the milk were not treated with rBST, but requiring that the label contain the additional statement that "no significant difference has been shown between milk derived from rBST-treated and non-rBST treated cows" (FDA 1994: 6280).

On January 18, 2001, the FDA issued draft guidelines for voluntary labeling of GM foods, which it now refers to as "bioengineered foods." The draft guidelines first reiterated that the agency had "no basis for concluding that bioengineered foods differ from other foods in any meaningful or uniform way, or that, as a class, foods developed by the new techniques present any different or greater safety concern than foods developed by traditional plant breeding" (FDA 2001: 4839). In the words of the statute, the fact that food contained GM materials was not "material" under section 201(n) of the FDCA.

The agency further noted, however, that section 403(j) of the Act requires that each food label "bear a common or usual name or, in the absence of such a name, an appropriately descriptive term" (FDA 2001). The agency read this requirement, in connection with the section 201(n) requirement that food labels reveal all "material" facts, to require labeling four situations:

> If a bioengineered food is significantly different from its traditional counterpart, such that the common or usual name no longer adequately describes the new food, the name must be changed to describe the difference.
>
> If an issue exists for the food or a constituent of the food regarding how the food is used or consequences of its use, a statement must be made on the labeling to describe the issue.
>
> If a bioengineered food has a significantly different nutritional property, its labeling must reflect the difference.
>
> If a new food includes an allergen that consumers would not expect to be present based on the name of the food, the presence of that allergen must be disclosed in the labeling. (FDA 2001)

Under the draft guidelines, labeling would be mandatory for foods and food products that fit into one of the four categories.

With respect to voluntarily labeling, the draft guidelines advised that a simple statement on the label that a food contains "genetically engineered material" or material that "was produced using biotechnology" would generally not be misleading, nor would statements accurately explaining the purpose of the genetic modification (e.g., "to increase yield") (FDA 2001).

The draft guidelines' advice on how a manufacturer might go about labeling foods that do not contain GM materials was less straightforward. First of all, "statements like 'not genetically modified' and 'GMO free,' that include the word 'modified' are not technically accurate unless they are clearly in a context that refers to bioengineering technology," because "'genetic modification' means the alteration of the genotype of a plant using any technique, new or traditional" (FDA 2001). That may be what the term "genetic modification" means to the scientists at the FDA, but ordinary people in the age of modern agricultural biotechnology almost certainly understand that the terms "GMO free" and "not genetically modified" refer to modern genetic engineering techniques, and not traditional plant breeding techniques. The FDA document admitted as much two sentences later when it noted that "consumers do not have a good understanding that essentially all food crops have been genetically modified and that bioengineering technology is only one of a number of technologies used to genetically modify crops" (FDA 2001). If consumers lack this understanding, it is probably because they understand that a label that refers to "GMO" or "genetic modification" refers to modern biotechnology and not traditional technologies.

Perhaps the most controversial advice in the draft guidelines was its strong suggestion that manufacturers avoid the words "free" and "zero." The document properly observed that it is often impossible, as a practical matter, to be absolutely certain that a particular piece of food has not been contaminated adventitiously due to poor segregation, cross-fertilization, or simple human error. The woeful StarLink episode is certainly evidence of the difficulties encountered in keeping an unwanted genetically engineered grain out of the grain distribution system (Bratspies 2003). In that remarkable failure of governmental oversight, human food supplies became contaminated by a potentially allergenic strain of GM corn that had been approved solely for use as animal feed. The draft guidelines recognized that, in the end, "it may be necessary to conclude that the accuracy of the term 'free' can only be ensured when there is a definition or threshold above which the term could not be used" (FDA 2001). The European Communities have, for example, established a threshold of 1% for the presence of GM material in food labeled "GMO free," but the FDA has so far taken the position that it lacks "information with which to establish a threshold level of bioengineered constituents or ingredients in foods for the statement 'free of bioengineered material'" (FDA 2001). The draft guidelines therefore advise that "the term 'free' either not be used in bioengineering label statements or that it be in a context that makes clear that a zero level of bioengineered material is not implied" (FDA 2001). Instead, the agency recommends that manufacturers employ awkward phrases like "We do not use ingredients that were produced using biotechnology" (FDA 2001). In any event, manufacturers claiming that a product was not "developed using bioengineered material" would bear the burden of substantiating that claim by either "validated testing" or by documenting "special handling" that was "appropriate to maintain segregation" of GM materials (FDA 2001).

Finally, the draft guidelines warned manufacturers that statements on food labels concerning the absence of GM materials should not in any way imply that such products were superior to products that do use such materials (FDA 2001). The agency wondered whether it would help to add a clarifying statement "that the absence of the use of bioengineering does not make the food superior to food not so labeled" (FDA 2001: 4840).

## THE CONSTITUTIONAL DEBATE

Whatever the outcome of the policy debate over the wisdom of labeling GM foods, any legal requirements imposed by the federal government must pass constitutional muster. A company that does not want to label GM foods may claim that the labeling requirement is inconsistent with its First Amendment right "not to speak." While there may have been a time when this argument would be dismissed as preposterous on its face, since a corporation is not a person but a legal entity created by the state to serve the interests of the public, it has a good deal of vitality in the wake of the remarkable Second Circuit opinion in *International Dairy Foods v. Amestoy* (1996).

In *Amestoy*, several dairy companies and industry associations challenged a Vermont statute requiring dairy manufacturers to identify products that were, or might have been, derived from dairy cows treated with rBST. The district court denied their request for a preliminary injunction. Concluding that "[t]he wrong done by the labeling law to the dairy manufacturers' constitutional right not to speak is a serious one that was not given proper weight by the district court," the court of appeals reversed and ordered the district court to issue the requested injunction (*International Dairy Foods v. Amestoy* 1996: 71).

The court held that the statute "contravene[d] core First Amendment values," because it "indisputably" required dairy producers "to speak when they would rather not." The court recognized that, under the relevant Supreme Court precedent, government may require companies to reveal aspects of their products to consumers, even if they would rather not, but the court found that the government in this instance did not have "a substantial interest" in doing so. The court noted that the State of Vermont did not claim that the labeling statute was intended to protect public health and safety (*International Dairy Foods v. Amestoy* 1996: 73). The governmental interest underlying the statute was merely the public's "right to know," and the court found that strong consumer interest was "insufficient to justify compromising protected constitutional rights" (73). The court noted that if consumer interest in hormone treatment were sufficient to justify a labeling requirement, there would be no end to the information that a state could make a company add to food labels. Absent "some indication that…information bears on a reasonable concern for human health or safety or some other sufficiently substantial governmental concern, the manufacturers cannot be compelled to disclose it" (73). The court was confident that the free market would ensure that companies would provide all of the information that the consumers really wanted, noting that "[t]hose consumers interested in such information should exercise the power of their purses by buying products from manufacturers who voluntarily reveal it" (74).

The *Amestoy* case has been the subject of much scholarly criticism, and there are good reasons to believe that it was wrongly decided (Beaudoin 1999; Marden 2003). The Supreme Court has explicitly stated that the Constitution "accords a lesser protection to commercial speech than to other constitutionally guaranteed expression" (*Central Hudson Gas & Elec. Corp. v. Public Serv. Comm'n* 1980), and state and federal governments have considerable leeway in requiring corporations engaged in commerce to make information available to consumers to maintain a "fair bargaining" process (*City of Cincinnati v. Discovery Network, Inc.* 1993), to prevent consumer deception (*Zauderer v. Office of Disciplinary Counsel* 1985), and to ensure that consumer decisions are "intelligent and well-informed" (*Virginia Board of Pharmacy v. Virginia Consumers Council* 1976). The Federal Trade Commission (FTC) has promulgated rules requiring labeling aimed at informing consumers without any serious constitutional objection (FTC, 2007). The Court has also recognized a distinction between mandatory disclosure requirements on foods and products and regulations preventing companies from communicating with consumers, noting that mandatory disclosure requirements "trench much more narrowly on [a corporation's] interests than do flat prohibitions on speech" (*Zauderer v. Office of Disciplinary Counsel* 1985). The dissenting judge in *Amestoy* concluded that the state's "worries about rBST's impact on human and animal health, fears for the survival of small dairy farms, and concerns about the manipulation of nature through biotechnology" were substantial enough to justify the very limited intrusion that a simple disclosure requirement would impose upon milk producers. Although the *Amestoy* majority expressed sympathy for Vermont consumers who wanted to know the contents of the milk they purchased at the store, it in fact afforded very little deference to the state legislature's determination that their interests were substantial and warranted protection.

Whether or not the *Amestoy* case is correctly decided, it should not stand in the way of reasonable labeling requirement for GM foods. The federal government has many legitimate reasons for requiring food manufacturers to disclose to consumers the fact that their products have been genetically modified. First, Congress or the FDA could legitimately conclude that the health consequences of human exposure to GM foods are not sufficiently well understood to support a finding that they may be safely consumed in all cases even if, by the manufacturer's estimation, they are generally recognized as safe. For example, so little is known about how to measure the allergenic potential of novel proteins in foods that Congress or the FDA could reasonably conclude that food labels should contain information about GM content to allow persons with allergies to make informed decisions about whether to consume them.

Second, Congress could legitimately conclude that the FDA is not the final arbiter of food safety and enact its own labeling requirement designed to allow consumers to make their own safety determinations. Consumers have good reason to distrust state and federal agencies that, under the watchful eye of the regulated industry, consistently find no health risks in GM foods. Although the State of Vermont in *Amestoy* did not assert a health and safety rationale for its labeling requirement, it easily could have relied upon the indirect effects of rBST use on the availability of critical antibiotics. The FDA was not persuaded that such indirect effects warranted a regulation banning the use of rBST; a consumer might disagree and decide not to encourage the overuse of antibiotics on milk cows by declining to drink milk from cows treated with rBST.

Third, the government has a legitimate interest in protecting citizens who, for moral reasons, want to avoid GM foods because they believe that it is not right for human beings to "play God" to that extent. Similarly, the government may have a legitimate interest in protecting the choices of citizens who practice religions that prescribe dietary restrictions forbidding consumption some or all GM foods. The government could similarly decide that labeling is an appropriate way to protect vegetarians from consuming vegetables containing proteins from animal genes. The court in *Amestoy* did not address such nonsafety concerns because the state did not invoke such concerns as a rationale for its rBST labeling requirement (Beaudoin 1999: 258).

## CONCLUSIONS

Despite the considerable efforts on the part of the biotechnology industry put the public's mind at ease about the safety and purity of their products, there appears to be an emerging worldwide consensus that GM foods should be labeled to allow consumers to choose whether or not to purchase such foods. A U.S.-E.U. Biotechnology Consultative Forum concluded in 2000 that "at the very least, the United States and the European Union should establish content-based mandatory labeling requirements for finished products containing novel genetic material" (U.S.-E.U. Biotechnology Consultative Forum 2000: 6). Even many observers who generally support greater availability of GM foods believe that the industry could eliminate a great deal of needless controversy by simply placing a discrete label on GM foods (Raeburn 2000).

Thus far, the relevant regulatory agencies have resisted implementing any form of mandatory labeling regime for food containing GM materials. Congress has over the past several years considered several bills that would require the federal government to implement a mandatory labeling program, but none of these bills has even emerged from the relevant committee (Galant 2005: 153). Given continued strong opposition by an economically and politically powerful industry, a mandatory labeling program is not likely to be implemented in the next several years, absent some crisis-inducing event, like another StarLink fiasco, with clear human health implications. Nevertheless, as labeling regimes begin to function in such places as the European Communities and their practicality is tested over time, mandatory labeling may still emerge as a serious regulatory possibility.

Should Congress or the FDA eventually decide to require mandatory labeling, four potential models exist for such a labeling regime. First, the government could require labeling only for foods for which the relevant agency has determined that the genetic modification is likely to cause adverse health effects or have a significant impact on nutrition. This is the approach that the FDA has adopted in its guidelines. Second, the government could require that all foods containing any detectable proteins resulting from genetic modifications be labeled. Third, the government could require labeling for all foods containing such detectable proteins in levels above a prescribed percentage. This appears to be the approach in place in the European Union. Fourth, the government could require that all foods that have been genetically modified and all foods directly or indirectly derived from GM foods be labeled. This approach, which is not in effect anywhere, would require labeling of oils and other products derived from GM plants and conceivably even meats derived from animals that have consumed GM feeds.

Although the constitutional legitimacy of GM food labeling requirements is not entirely free from doubt, a federal agency should be able to articulate a rationale, based upon a "substantial" governmental interest, capable of surviving the relatively modest judicial scrutiny afforded to such limited intrusions on commercial speech under the relevant Supreme Court First Amendment precedent. It is unlikely that the U.S. Constitution is so solicitous of the commercial interests of its corporate citizens that it prevents the U.S. government from requiring companies to provide to its living citizens the same information that is readily available to the citizens of most European countries.

Whether or not Congress or the FDA decides to require mandatory labeling, some degree of voluntary labeling is likely to be implemented by companies seeking to appeal to a growing market for pure and otherwise "natural" foods.

Although voluntary labeling poses a very real threat to consumer well-being if companies are allowed to place misleading labels on foods that either do contain GM materials or do not contain any constituents for which GM plant varieties have been developed, the FDA should not adopt a pedantic technical approach to labeling restrictions. The public will not be misled by the term "GMO free" to believe that food so labeled does not contain materials from plants that have been modified through traditional breeding technologies.

Finally, the FDA should not be overly solicitous of food industry fears that consumers will mistakenly conclude from labeling that food that contains little or no GM materials is safer or more wholesome than food that does contain such materials. Many consumers will no doubt select food that contains little or no GM materials precisely because of their belief that it is safer and/or more wholesome. Although GM labeling proponents cannot conclusively prove that GM foods are generally less safe or wholesome, GM labeling opponents have not proved that they are safe and wholesome. Surprisingly little direct testing has been done with GM foods, especially in the critical area of allergenicity. At most, the proponents can say that most scientists believe that the two appear to be substantially equivalent for public health purposes, even though they have been consciously designed to be substantially different for one or more commercial purposes. Overbearing governmentally imposed restrictions on reasonably accurate food labeling will deprive consumers who want to know whether their food contains GM materials of the very information that they desire and should, in a market-based food distribution system, have at their fingertips as they travel the grocery store aisles.

## References

*Alliance for Bio-Integrity v. Shalala*, 116 F. Supp. 2d 166 (D.D.C. 2000).

Barrett, William P. December 27, 1999. Food-Label Follies. *Forbes*, 30.

Beales, J. Howard. 2000. Modification and Consumer Information: Modern Biotechnology and the Regulation of Information. *Food Drug L.J.* 55:105.

Beaudoin, Kirsten S. 1999. On Tonight's Menu: Toasted Cornbread with Firefly Genes? *Marq. L. Rev.* 83:237.

Bohrer, Robert A. 1994. Food Products Affected by Biotechnology. *U. Pitt. L. Rev.* 55:653.

Bratspies, Rebecca M. 2003. Myths of Voluntary Compliance: Lessons from the StarLink Corn Fiasco. *Wm. & Mary Envtl. L. & Pol'y Rev.* 27:593.

Burk, Dan L. 1997. Milk Free Zone: Federal and Local Interests in Regulating Recombinant BST. *Colum. J. Envtl. L.* 22:227.

Carter, Colin A., & Guillaume P. Gruere. 2003. Mandatory Labeling of Genetically Modified Foods: Does It Really Provide Consumer Choice? *AgBioForum* 6:68.

Central Hudson Gas & Elec. Corp. v. Public Serv. Comm'n, 447 U.S. 557, 562–563 (1980).

*City of Cincinnati v. Discovery Network, Inc.*, 507 U.S. 410, 426 (1993).

Degnan, Fred H. 2000. Biotechnology and the Food Label: A Legal Perspective. *Food Drug L.J.* 55:301.

Endres, A. Bryan. 2005. Revising Seed Purity Laws to Account for the Advantitious Presence of Genetically Modified Varieties: A First Step Towards Coexistence. *J. Food Law & Pol'y* 1:131.

Engel, Karl-Heinz, Gary R. Takeoka, & Roy Teranishi. 1995. Foods and Food Ingredients Produced via Recombinant DNA Techniques. In *Genetically Modified Foods: Safety Issues,* ed. Karl-Heinz Engel, Gary R. Takeoka, & Roy Teranishi. American Chemical Society, Symposium Series No. 605. Washington, D.C.: American Chemical Society, 2.

FDA (Food and Drug Administration). 1986. Irradiation in the Production, Processing and Handling of Food. *Fed. Reg.* 51:13376.

FDA (Food and Drug Administration). 1992. Statement of Policy: Foods Derived from New Plant Varieties. *Fed. Reg.* 57:22984.

FDA (Food and Drug Administration). 1993a. Food Labeling: Mandatory Status of Nutrition Labeling and Nutrient Content Revision, Format for Nutrition Label. *Fed. Reg.* 58:2079.

FDA (Food and Drug Administration). 1993b. Food Labeling; Foods Derived from New Plant Varieties. *Fed. Reg.* 58:25837.

FDA (Food and Drug Administration). 1994. Interim Guidance on the Voluntary Labeling of Milk and Milk Products from Cows That Have Not Been Treated with Recombinant Bovine Somatotropin. *Fed. Reg.* 59:6279.

FDA (Food and Drug Administration). 2001. Draft Guidance for Industry: Voluntary Labeling Indicating Whether Foods Have or Have Not Been Developed Using Bioengineering. *Fed. Reg.* 66:4839. Available at http://www.cfsan.fda.gov/~ dms/biolabgu.html.

Food, Drug and Cosmetics Act of 1938, as amended, 21 U.S.C. §§ 341, et. seq.

Frank, Richard. November 30, 1999. Remarks. In *Public Meeting on Biotechnology in the Year 2000 and Beyond,* ed.U.S. Food and Drug Administration. Available at http://www.fda.gov/ohrms/dockets/dockets/99n4282/99n4282tr.htm.

FTC (Federal Trade Commission). 2007. Care Labeling of Textile Wearing Apparel and Certain Piece Goods. Code of Fed. Reg. 16:§ 423.

Galant, Carl R. 2005. Labeling Limbo: Why Genetically Modified Foods Continue to Duck Mandatory Disclosure. *Hous. L. Rev.* 42:125.

Goldman, Karen A. 2000. Labeling of Genetically Modified Foods: Legal and Scientific Issues. *Geo. Int'l Envtl. L. Rev.* 12:717.

Grossman, Margaret Rosso. 2005. Traceability and Labeling of Genetically Modified Crops, Food, and Feed in the European Union. *J. Food L. Pol.* 1:43.

Groth, Edward. November 18, 1999. Remarks. In *Public Meeting on Biotechnology in the Year 2000 and Beyond,* ed. U.S. Food and Drug Administration. Available at http://www.fda.gov/ohrms/dockets/dockets/99n4282/99n4282tr.htm.

Hoban, Thomas J. 1998. International Acceptance of Agricultural Biotechnology. In *Agricultural Biotechnology and Environmental Quality: Gene Escape and Pest Resistance,* ed. Ralph W.E. Hardy & Jane Baker Segelken. National Agricultural Biotechnology Council, Rept. 10. Ithaca, N.Y.: National Agricultural Biotechnology Council, 59.

Howle, David & Brian Sim. 1998. Regulatory and Economic Aspects of Accessing International Markets Workshop. In *Resource Management in Challenged Environments,* ed. Ralph W.E. Hardy, Jane Baker Segelken, & Monica Voionmaa. National Agricultural Biotechnology Council, Rept. 9. Ithaca, N.Y.: National Agricultural Biotechnology Council, 23.

Industry Opposes Biotech Labeling. February 28, 2000. *Chemical Market Reporter,* 5.

International Dairy Foods v. Amestoy, 92 F.3d 67 (2d Cir. 1996).

Jacobson. November 18, 1999. Remarks. In *Public Meeting on Biotechnology in the Year 2000 and Beyond,* ed. U.S. Food and Drug Administration. Available at http://www.fda.gov/ohrms/dockets/dockets/99n4282/99n4282tr.htm.

Keith, Kendal. November 30, 1999. Remarks. In *Public Meeting on Biotechnology in the Year 2000 and Beyond,* ed. U.S. Food and Drug Administration. Available at http://www.fda.gov/ohrms/dockets/dockets/99n4282/99n4282tr.htm.

Kirby, Sarah L. 2001. Genetically Modified Foods: More Reasons to Label Than Not. *Drake J. Agric. L.* 6:351.

Kysar, Douglas A. 2001. Sustainability, Distribution, and the Macroeconomic Analysis of Law. *Boston Coll. L. Rev.* 43:1.

Lappe, Marc, & Britt Bailey. 1998. *Against the Grain: Biotechnology and the Corporate Takeover of Your Food.* Monroe, Maine: Common Courage Press.

Leggio, Kelly A. 2001. Limitations on the Consumer's Right to Know: Settling the Debate over Labeling of Genetically Modified Foods in the United States. *San Diego L. Rev.* 38:893.

Marden, Emily. 2003. Risk and Regulation: U.S. Regulatory Policy on Genetically Modified Food and Agriculture. *B.C. L. Rev.* 44:733.

Mayer, James, & Michelle Cole. November 6, 2002. Measures Labeling Altered Food Contents, Health Care Fail to Get Support. *Oregonian,* A1.

McGarity, Thomas O. 1987. Federal Regulation of Agricultural Biotechnologies. *U. Mich. J. L. Ref.* 20:1089.

McGarity, Thomas O. 2002. Seeds of Distrust: Federal Regulation of Genetically Modified Foods. *U. Mich. J. L. Ref.* 35:403.

McGarity, Thomas O., Sindney A. Shapiro, & David Boller. 2004. *Sophisticated Sabotage.* Washington, D.C.: Environmental Law Institute.

Miskiel, Frank J. 2001. Voluntary Labeling of Bioengineered Food: Cognitive Dissonance in the Law, Science, and Public Policy. *Cal. W. L. Rev.* 38:223.

National Academy of Sciences. 2000. *Genetically Modified Pest-Protected Plants: Science and Regulation.* Washington, D.C.: National Academies Press.

Nutrition Labeling and Education Act of 1990, Pub. L. 101–535, 104 Stat. 2353 (1990), 21 U.S.C. § 343(q).

Organic Foods Production Act of 1990, 7 U.S.C. 6501–6522.

Raeburn, Paul. April 17, 2000. Biotech Foods Aren't Out of the Woods Yet. *Business Week,* 56.

Royal Society of Canada. 2001. *Elements of Precaution: Recommendations for the Regulation of Food biotechnology in Canada: Expert Panel Report of the Future of Food Biotechnology.* Ottawa, Ontario: Royal Society of Canada.

Seeds of Change: In the U.S. and Elsewhere, the Food Supply Is Being Genetically Altered. September 1999. *Consumer Reports,* 23.

Sunstein, Cass R. 2003. Beyond the Precautionary Principle. *U. Pa. L. Rev.* 151:1003.

U.S.-E.U. Biotechnology Consultative Forum. 2000. *Final Report.* Available at http://italy.usembassy.gov/pdf/other/biotech.pdf.

USDA (United States Department of Agriculture). December 21, 2000. National Organic Program. Fed. Reg. 65:80548–80549.

*Virginia Board of Pharmacy v. Virginia Consumers Council,* 425 U.S. 748, 765 (1976).

*Zauderer v. Office of Disciplinary Counsel,* 471 U.S. 626 (1985).

Zeichner, Lauren. 2004. Product vs. Process: Two Labeling Regimes for Genetically Engineered Foods and How They Relate to Consumer Preference. *Environs Envtl. L. Pol'y J.* 27:467.

9

# Regulatory Barriers to Consumer Information about Genetically Modified Foods

*Philip G. Peters & Thomas A. Lambert*

Genetic modification, or "bioengineering," of plants and animals presents an intriguing problem for regulators. On the one hand, genetic modification promises substantial improvements in agricultural efficiency, nutritional content of foods, and environmental protection. On the other hand, regulation of genetically modified (GM) foods may be justified by the potential externalities involved in GM food production, the risks associated with consumption of GM foods, or the desire to ensure consumer autonomy over consumption decisions. These externality-based, autonomy-based, and risk-based concerns give rise to both utilitarian and rights-based arguments for regulatory intervention.

In light of bioengineering's potential social benefits, most actual and proposed regulation of GM foods has not directly restricted production but has instead focused on informing consumers of GM content (Beales 2000). Policy makers and commentators thus seem to have discounted the externality-based arguments for regulation, which would seem to support production restrictions, and have focused instead on risk-based and autonomy-based concerns, which (unlike externalities) may be redressed via labeling.

In this chapter, we assume, in concert with the majority of mainstream scientists that have considered the issue, that GM foods currently on the market are substantially equivalent to their non-GM counterparts in terms of proven health risks (Preston 2005). If that is indeed the case, and if, as the majority of adopted and proposed regulatory regimes suggest, externality-based concerns are not sufficient to justify production restrictions, what is the optimal approach to the regulation of GM foods? More specifically, if an autonomy-enhancing information-provision scheme is the appropriate regulatory response, how should the informational regulation be structured? We conclude that some sort of voluntary negative labeling scheme would represent the optimal regulatory response but that the legal status quo in the United States falls short of the ideal because it erects unjustifiable legal barriers to the voluntary provision of desirable information regarding the bioengineered status of non-GM foods. We thus advocate removal of those legal barriers.

## WHAT LABELING OPTIONS ARE AVAILABLE, AND WHICH IS OPTIMAL?

Labeling is, of course, a fairly nonrestrictive means of regulating GM foods. Policy makers could have outright banned the production and/or marketing of such foods, just as the notorious Delaney Clause once did for foods known to contain even trace amounts of carcinogens. Alternatively, policy makers could have sought to prevent gene flow and other externalities associated with GM food production by directly regulating the means of production (e.g., by requiring buffer zones between GM and non-GM crops). For the most part, however, policy makers around the world have eschewed these regulatory options and have concentrated instead on the question of how product labeling should be used to inform consumers of genetic modification. They have therefore generally focused on the right half of the scale shown in figure 9.1: debating whether labeling of GM products should be mandatory and, if not, what stance the government should take toward voluntary labeling efforts. In this chapter section, we evaluate the merits of the regulatory options that have been adopted or seriously considered (i.e., those on the right side of the scale).

### Mandatory Labeling

The most intrusive labeling regime, versions of which have been adopted by, among others, the European Union, Japan, South Korea, Australia, and New Zealand, mandates the labeling of GM foods and those containing GM ingredients. In certain circumstances, the regimes permit "May Contain" notices, which state merely that a GM ingredient may have been used in preparing the food product at issue (e.g., "The corn syrup in this product was derived from corn that may have been bioengineered"). In general, however, they require definitive "Does Contain" notices stating that the product actually contains a GM ingredient (e.g., "The corn syrup in this product was derived from bioengineered corn"). The latter notices obviously provide more information to consumers but also impose significantly higher costs on the seller, who must trace GM ingredients throughout the

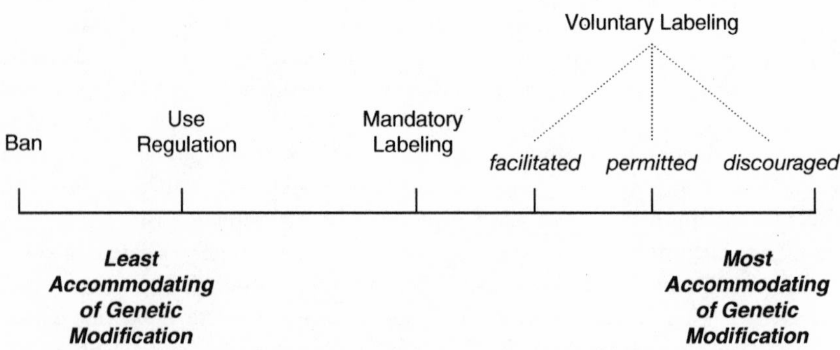


Figure 9.1. Options for regulating GM foods.

production process, segregate GM from non-GM ingredients and products, and document such segregation and tracing (Beales 2000).

Policy makers should therefore consider whether those significant costs are justified. At present, they probably are not. There is no externality-based rational for a mandatory labeling regime, for the labeling of GM food will do little or nothing to alleviate the potential cost spillovers involved in the production or marketing of GM seeds (e.g., gene flow). Advocates of mandatory labeling have therefore sought to articulate risk-based and autonomy-based rationales. Upon reflection, however, neither set of rationales will currently justify a mandatory labeling regime. Because a voluntary labeling regime has concrete advantages, policy makers should remove the barriers that currently impair voluntary labeling before concluding that mandatory labels are needed.

## Risk-Based Justifications

With respect to risk, mandatory labeling advocates have primarily argued that labeling should be required because of potential toxicity (Hadfield and Thomson 1991), or because incorporated genetic material may transfer allergenicity, as when genetic material from a Brazil nut, to which many humans are allergic, is incorporated into a bioengineered potato (Marden 2003; Thue-Vasquez 2000). In addition, some mandatory labeling advocates have invoked the so-called Precautionary Principle, which counsels regulatory precaution in the face of potential but uncertain risks, because GM foods do not have a long track history of safe use.

These risk-based rationales for mandatory labeling largely fall away if we assume, as the majority of scientists seem to believe, that the risks associated with consumption of currently available GM foods are insignificant (Preston 2005). Moreover, even accepting that there are discrete risks associated with the consumption of some GM foods, such as when a bioengineered food incorporates an allergen from an organism from which genetic material was transferred, the information that is relevant to affected consumers is not the fact that the product has been bioengineered but is instead that it may pose the particular risk at issue. Thus, a requirement that all GM foods be labeled as such is, at best, a blunt tool for averting any discrete risks associated with consumption of particular bioengineered foods. If the motivating concern is risk reduction, any mandated label should state the actual risk at issue (e.g., potential allergenicity), not the fact that the food was produced using a particular process.

Mandatory labeling advocates might retort, of course, that the long-term risks associated with GM food consumption may be currently unknown and thus impossible to disclose with specificity. But that is true with literally every new food ingredient and production process. Indeed, this very argument was asserted a century ago with regard to conventional hybridization (Riddick et al. 1997). If the unknown risks associated with new ingredients and/or production processes could justify mandatory labeling of products containing those ingredients or manufactured using those processes—even when science had failed to identify any significant risks associated with the ingredients or processes—then consumers would be overwhelmed with labeling detail.

More fundamentally, it is not clear what mandatory labels do to alleviate potential but uncertain risks. Obviously, mandatory labeling would not actually reduce any such risks; it would merely assist risk-averse individuals in avoiding unknown risks by forgoing consumption of labeled products. As we explain in more detail below, if there are enough risk-averse consumers, sellers of products that do not involve the unknown risks at issue will identify their products as not presenting the unknown risk (e.g., sellers could label their products as "non-GM"), so *mandatory* labeling would not be necessary. Absent a critical mass of risk-averse consumers, such voluntary labeling might not occur, but in that case it would seem inappropriate to force all sellers of GM foods to shoulder the burden of costly labeling in order to cater to the risk preferences of only a few consumers.

## Autonomy-Based Justifications

Autonomy-based rationales for mandatory labeling have typically relied on consumers' purported "right to know" whether the food they are consuming contains GM ingredients or on survey data suggesting that a substantial majority of consumers desire mandatory labeling of GM foods. In addition, mandatory labeling advocates have argued that requiring labels on GM foods is the best way to accommodate the heterogeneous preferences of individuals who may prefer to purchase non-GM foods, despite the apparent lack of any evidence that they offer health benefits superior to their GM counterparts. None of these arguments is convincing.

*Consumers' Right to Know:* While talk of consumers' "right to know" makes for effective rhetoric, there are conceptual problems involved in recognizing such a right. It would be difficult to conceive of such a right as part of a scheme of natural rights, for the right lacks logical boundaries and ultimately restricts liberty in a manner not necessary to protect anyone's person or property. First, consider the "boundaryless" nature of the purported right. As Howard Beales (2000) has explained:

> A "right to know" could be invoked to justify labeling about any detail of the production process, from use of chemical fertilizers, to the wage rate and national origin of the workers who planted and harvested the crop, to the labor practices of the manufacturer, to the soil conservation practices of the farmer. It is impossible to list all the things that might matter to everyone. (p. 109)

Because there is no principled way to cabin the consumer's so-called right to know, the right seems to be a poor candidate for inclusion in any theory of natural rights.

More important, the right restricts liberty by precluding voluntary transactions that do not injure the person or property of third parties. Because the purported right to information occurs as part of a voluntary transaction (i.e., a purchase contract between the seller of the GM food and the consumer), the level of information provided by the seller is a function of the bargain between the parties. If consumers want more information, they can demand it, and sellers can decide whether to provide the information, refuse to provide the information, or provide the information only in exchange for additional consideration (i.e., a higher purchase price). To say that there is a "right" to this sort of information is to say that buyers and sellers are forbidden to enter a transaction where the information is not provided as part of the deal. Thus, the purported "right to know" is ultimately liberty restricting, not liberty enhancing, for the parties are precluded from entering a transaction that they otherwise could have entered. Moreover, because the consumer desiring information regarding bioengineered status could obtain that information through market processes, the liberty restriction is unnecessary to satisfy others' needs—or even their wants. Natural rights status, then, seems improper.

Of course, lawmakers could create a *positive* right to know by forbidding GM food sales unaccompanied by information regarding the food product's bioengineered status. Creation of such a right would seem to require some policy justification, however, for the right involves a significant "downside" in that it ultimately prevents consumers who do not value the information at issue from purchasing bioengineered food products at a lower price but without the mandated information. A frequently cited justification for mandating GM labels (thereby creating a positive right to information) is that consumers have a right to know that the food they are purchasing has been bioengineered. But that justification is obviously circular. Accordingly, mandatory labeling must be justified on grounds other than the "right to know." It might, for example, be justified by the fact that many citizens appear to desire mandatory labeling or by respect for consumers' heterogeneous preferences. Considered closely, however, each of these alternative justifications fails.

*Democratic Principles:* In public opinion polls, substantial majorities of respondents consistently express a desire for mandatory labeling of GM food products. For example, the Pew Initiative on Food and Biotechnology, which since March 2001 has conducted regular surveys of Americans as to their views of GM food and its regulation, has consistently found that Americans overwhelmingly favor labeling of GM food products (Pew Initiative on Food and Biotechnology 2001). In the October 2004 Pew survey, 92% of respondents supported or strongly supported "labeling all food that is genetically modified," and 91% supported or strongly supported "labeling all processed food that contains genetically modified ingredients" (Pew Initiative on Food and Biotechnology 2004). In a recent survey by researchers from the University of Maine and Ohio State University, 85% of respondents indicated that they wanted GM foods to be labeled, and of those respondents, 88% indicated that the labeling should be mandatory (Teisl et al. 2003). Other public opinion surveys have generated similar results (Fisher 2002).

A number of commentators argue that these polling data justify mandatory labeling (Streiffer and Rubel 2004). For example, observing that "Congress has... an obligation to act as a representative of the people on certain kinds of decisions," Philosophers Robert Streiffer and Alan Rubel (2003) contend that, "[g]iven the persistent and overwhelming majority of Americans who think that there should be mandatory labeling on GM products, a Congressional decision not to require labeling would violate that obligation, regardless of whether doing so leads to fewer GM products on the market" (page 142).

Polling data ultimately provide a weak justification for mandatory labeling, however. First, the data are untrustworthy. Most of the surveys cited in support of a mandatory labeling policy simply ask consumers whether or not they desire mandatory labeling of GM food (or, more commonly, whether they "support" or "strongly support" mandatory labeling). Absent some mention of any downside to such labeling, one would expect the vast majority of consumers—rationally preferring more information to less information, all else being equal—to support such labeling. The data would likely be quite different if respondents were confronted with one or more of the inevitable downsides of mandatory labeling, such as price increases, higher taxes, or reduced consumer choice. Survey respondents substantially moderate their opposition to GM foods when confronted with the tradeoffs involved in eliminating bioengineering (Noussair et al. 2004), and one would similarly expect enthusiasm for mandatory labeling to wane if the costs of such labeling were brought "on screen." Because salient cost information has generally been omitted from public opinion surveys regarding mandatory labeling, survey data are somewhat unreliable.

Even if the polling data were not suspect, however, opinion surveys would provide a weak basis for mandating labeling, for such "policy by polling" embodies a conception of democracy that is both undesirable and inconsistent with constitutional design. The view that lawmakers should require GM labeling because most citizens want it mandated stems from a "pluralist" conception of democracy, in which the role of elected officials (and the unelected bureaucrats under their control) is to maximize the preferences of individual citizens. In a "republican" democracy, by contrast, the role of elected officials is more reflective; they are charged with deliberating over expressed citizen desires and selecting policies that, after consideration, appear to maximize public welfare (Sunstein 1988). Under this more deliberative form of democracy, which appears to be more desirable as a policy matter and more consistent with the intent of the framers, the mere fact that majorities of citizens want mandatory labeling of GM foods would not be enough to justify a mandatory labeling law. Instead, elected officials should take survey data into account (recognizing the limits of such data) but then select the policy outcome most likely to enhance public welfare in the long run (Miller 2003). For reasons stated below, a policy that encouraged (or at least did not discourage) voluntary labeling of non-GM foods would be most likely to maximize citizen welfare.

*Accommodation of Heterogeneous Preferences:* Perhaps the strongest argument in favor of mandatory labeling is that such labeling would accommodate consumers' heterogeneous preferences. Even if the consensus of mainstream science is that GM foods are generally safe to consume, many consumers may prefer to avoid such foods. Consumers may, for example, have different attitudes toward risk than the majority of scientists (i.e., they may be more risk averse), or they may oppose bioengineering for reasons unrelated to consumption risk (e.g., because they do not want to contribute to the potential externalities involved in cultivating GM crops or they want to support small local farmers who do not engage in bioengineering). If bioengineered foods must be labeled as such, consumers who prefer not to purchase such foods may exercise their preferences. Such an outcome is desirable on both utilitarian grounds (because utility is increased as more consumers are able to engage in precisely the trades they find most desirable) and non-utilitarian, autonomy-based grounds (because fewer consumers are duped or effectively coerced into supporting production processes they oppose). Accordingly, a number of commentators have argued for mandatory labeling of GM food products on grounds that such labeling best honors divergent consumer preferences (Kysar 2004; Streiffer and Rubel 2003).

*Mandatory Labeling Not Necessary:* This argument assumes, though, that the information necessary to accommodate heterogeneous consumer preferences will not be produced unless its provision is mandated. If there is reason to believe that such information will be made available absent government mandate, then the case for mandatory labeling is significantly weakened. If, by contrast, there is some reason to believe that the information will not be privately produced, then the case for mandatory labeling is strengthened. In fact, both theory and empirical evidence suggest that government mandate is not necessary to produce the sort of information necessary to accommodate diverse consumer preferences regarding GM foods.

First consider theory. Mandatory information provision may be appropriate when there is some market failure that is likely to prevent private production of the information at issue. Given the "public good" nature of information (i.e., it is easily transferable and can be consumed without being depleted), it frequently exhibits positive externalities (i.e., benefits that cannot be captured by the producer). If too much of the information's benefit is expropriated, so that the producer is not able to recoup his investment in the production effort plus a reasonable return, the producer will not create the information—even if the total benefits of the information exceed the costs of production (Stigler 1961). Thus, the government may appropriately undertake efforts, such as public research or mandatory disclosure, to ensure an optimal amount of information production when it is unlikely that private parties could extract enough gain from the information at issue to justify their efforts in creating it (Lambert 2004).

With respect to information regarding the bioengineered content of food products, however, there is no reason to doubt that private parties would be adequately motivated to provide the information necessary to permit consumers to choose between GM and non-GM foods. An impressive body of empirical evidence indicates that substantial percentages of food consumers in advanced economies are willing to pay sizable premiums for non-GM foods (Baker and Burnham 2001; Chern et al. 2002; McCluskey et al. 2001; Mendenhall and Evenson 2002; Noussair et al. 2004; Rousu et al. 2004; Tegene et al. 2003). Thus, food sellers have strong incentives both to offer non-GM products and to identify those products as being free of bioengineered ingredients. In the absence of government intervention, one would therefore expect voluntary use of "negative" labels (i.e., labels stating that a food product has not been produced using GM products), by which consumers would become informed as to which food products have and have not been bioengineered.

Empirical evidence (albeit anecdotal) confirms this expectation. In the last several years, numerous food sellers have begun labeling their non-GM food products in response to consumers' willingness to pay premiums for non-bioengineered foods. Such voluntary negative labeling has occurred *despite* the existence of substantial legal obstacles to identifying non-bioengineered food products as such. Removal of those legal obstacles, which Part III discusses in detail, would likely proliferate the marketing of non-GM foods in response to consumer demand, thereby enhancing both societal wealth and consumer autonomy. Thus, it is not necessary to require labeling in order to accommodate heterogeneous consumer preferences.

*Mandatory Labeling Not Optimal:* Nor is mandatory positive labeling the optimal means of providing consumer information. It is, in fact, inferior to voluntary negative labeling on at least four grounds. First, it is likely to involve higher administrative costs than a voluntary negative labeling scheme. Under either labeling regime, silence—i.e., the absence of a label—will convey some meaning: If positive labeling is mandated, the absence of a label will indicate that the food product is non-bioengineered; if labeling is voluntary (and, as the evidence suggests, enhanced willingness to pay creates an incentive for sellers to identify their non-GM products), the absence of a negative label will suggest that the product at issue is bioengineered. Given that the background legal rule will determine the meaning of silence, policy makers may minimize the costs of informing consumers by defining silence, which is of course a cheaper signal than labeling, so that it indicates the more common bioengineered status. In other words, if most food products contain GM products, then costs would be minimized by making silence signify, "Contains GM products"; if most foods do not contain GM products, then the cost-minimizing rule would define silence to mean, "Free of GM products." Because an estimated 70% of food on the shelves of American grocery stores contains GM ingredients (Mandel 2004), the costs of informing consumers of the bioengineered status of food products will be minimized by adopting a negative labeling regime under which the absence of a label signifies that the food contains some bioengineered product. And, as explained above, negative labels need not be mandated, for sellers have an incentive to provide such labels voluntarily.

Voluntary negative labeling is also superior to mandatory labeling on fairness/distributional grounds. While many consumers care about genetic modification and would pay at least some premium to avoid bioengineered foods, many consumers are indifferent to a food product's bioengineered status. From a fairness standpoint, it seems more appropriate to allocate the costs of informing consumers regarding bioengineered status to those consumers who actually value the information—that is, those who are willing to pay at least some premium for non-GM foods. Under a system of mandatory positive labeling, the consumers of GM food, who are least likely to value the information at issue, ultimately bear the cost of labeling. By contrast, a system of negative labeling distributes the costs of informing consumers to the consumers most likely to value information regarding bioengineered status—that is, purchasers of non-GM products.

Third, mandatory positive labeling may actually increase health risks by placing a governmental imprimatur on, and thereby exacerbating, irrational fears that GM foods pose significant health risks. Individuals, cognitive psychologists have discovered, are subject to heuristics that cause them to amplify the risks associated with unnatural occurrences (i.e., processes that involve "tampering with nature") (Sunstein 2004; Rozin 2001) and to give greater credit to more alarmist accounts of circumstances (Viscusi 1997). These mental shortcuts create a bias toward acceptance of frightening accounts of the health risks associated with so-called "Frankenfoods." That bias, then, tends to be confirmed by news media inclined toward sensationalism and by social forces that lead individuals to accept propositions simply because they believe others accept them (Kuran and Sunstein 1999). The bias will be further exacerbated if GM labeling is required, for any mandatory labeling law will inevitably convey the message that bioengineered status is, at the very least, important (otherwise, why would law makers have required the labeling?). And because labels are generally mandated only when the product being labeled is somehow harmful or less desirable as a result of the labeled attribute, a mandatory labeling law is also likely to send the message that genetic modification is harmful or bad. In short, mandatory labeling puts a government-sanctioned stamp of approval on individuals' irrational risk judgments, thereby exacerbating the mistaken perception that GM foods pose substantial consumption risks.

As individuals decide to forgo GM foods in order to avoid risks they perceive to be substantial, they will inevitably incur other risks. For example, cutting back on fruit, vegetable, and grain consumption may increase one's cancer risk, and deciding to pay more for non-GM foods will reduce the amount one could spend on other forms of risk reduction (e.g., health care). By exacerbating the preexisting biases against genetic modification, mandatory labeling may therefore cause a net increase in health risks by generating undesirable "risk–risk tradeoffs" in which the real risk incurred is substantially greater than the perceived risk avoided. A dramatic example of this process occurred when Zambian officials, apparently influenced by European laws mandating the labeling of GM foods, concluded that such foods could not be safely consumed and therefore refused shipments of American food aid, thereby placing thousands of impoverished Zambian citizens at risk of starvation (Cauvin 2002).

Finally, if mandatory labeling tends to strengthen consumers' incorrect perception that consumption of GM foods poses substantial health risks, it may ultimately reduce consumer choice. Colin Carter and Guillaume Gruère (2003), explaining why mandatory labeling has this effect, observe that food producers' profits are a function of the number of units they can sell and the profitability per unit. Per unit profits may be higher if GM ingredients are used (since such products are generally available at lower cost), but if use of such ingredients sufficiently reduces consumer demand (because consumers perceive the ingredients to be risky), then producers will steer clear of such inputs despite the cost advantages they offer. To the extent that mandatory labeling exacerbates the perception of riskiness by placing a governmental imprimatur on that perception, it may reduce the number of producers willing to use GM products and thus may ultimately reduce consumer choice. Indeed, a number of studies have suggested that mandatory labeling tends to drive GM products from grocery store shelves. For example, Kalaitzandonakes and Bijman (2003) and Bernauer and

Meins (2001) both found that GM products virtually disappeared from retail food outlets in Europe following enforcement of the E.U. mandatory labeling law. Others have found that GM foods are difficult to find in Japan, Australia, and New Zealand, which similarly mandate labeling of bioengineered foods (Carter and Gruère 2003).

## Voluntary Labeling

The discussion so far has focused on the distinction between mandatory labeling and voluntary labeling and has argued for the general proposition that a voluntary negative labeling scheme would best inform consumers of a food product's bioengineered status. Unlike mandated labeling, voluntary labeling allocates the cost of information to the individuals who want it. Market-driven voluntary labeling is also more flexible than mandated labels because voluntary labels can change more quickly as new products emerge, new risks are identified, and new data are obtained about old products. In addition, voluntary labeling permits sellers to appeal to the preferences of consumers who prefer GM-free foods for reasons that fall outside the narrow confines of the laws that currently govern mandatory labeling. Those laws have two very important limitations. First, the general labeling authority of the U.S. Food and Drug Administration (FDA) applies only when a food poses a risk to human safety. The FDA is not authorized to mandate labels that help consumers advance other goals, such as buying products that are produced in ways that reduce the risk of ecological harm. Second, the FDA interprets its authority over food labeling to require an all-or-nothing decision about acceptable risk. As a consequence, the FDA will not use its labeling authority to provide the information needed by consumers whose tolerance for uncertain risks is lower than the FDA's. Voluntary labeling fills that gap.

Our preference for voluntary labeling is premised on a regulatory regime that permits and even encourages sellers to respond to the varying preferences of consumers regarding some or all GM foods. We therefore do not intend to defend the voluntary labeling scheme that currently exists in the United States. For the reasons stated further below, we believe that the American status quo erects undesirable barriers to the dissemination of information regarding the bioengineered status of food products. Before addressing the deficiencies in U.S. law, however, we briefly distinguish the available voluntary labeling approaches, which vary in the degree to which the government facilitates private labeling efforts. The three main approaches are exemplified by the labeling regimes applicable to organic foods, kosher foods, and milk from cows that have not been treated with recombinant bovine somatotropin (rBST), a controversial growth hormone.

### The Organic Approach: Government Facilitates Voluntary Labeling

Under the most "pro-labeling" voluntary regimes, the government actively assists private labeling efforts by setting standards that define what certain labels or labeling terms will mean. This is the approach U.S. law has taken toward organic foods. As is the case with non-GM foods, no scientific evidence demonstrates that organic foods offer superior nutritional or food safety benefits over their conventionally grown counterparts (Institute of Food Technologists 2000). Accordingly, the FDA initially deemed some organic labels to be misleading in that they implied that the labeled food was superior to similar nonorganic food. Private organic certification agencies nonetheless proliferated. Those organizations worked on different geographic scales, catered to diverse sectors of the organic community, and applied divergent standards for what constitutes "organic" means of production. As a result, the significance of an organic label could vary, depending on which organization's standard was being applied. In 1990, Congress sought to put an end to this confusing situation and to honor consumer preferences for information regarding organic status by providing for the creation of uniform organic standards. The Organic Food Production Standards Act of 1990 required the U.S. Department of Agriculture (USDA) to develop national organic standards and establish an organic certification program based on recommendations from an expert panel, and final standards were finally adopted in 2002. By providing food producers and consumers with these standards, Congress and the USDA have facilitated the voluntary labeling of organic foods, thereby encouraging market segmentation that conforms to consumer preferences.

Recent experience with organic labels suggests, however, that government standard setting is not a labeling panacea. As noted, the USDA took 12 years to accomplish Congress's directive. The USDA initially proposed standards in December 1997, but its proposal was fiercely derided by the organic community. After receiving more than 275,000 written comments (more than it had ever received on any proposed rule), the department withdrew its proposed standards and started over. In March 2000, it proposed the alternative standards that were finally adopted on October 21, 2002 (National Organic Program, 2002) Still, however, many in the organic community maintain that the government-provided standards are inadequate. Believing the standards fail to distinguish production practices many consumers care about on ethical grounds (e.g., how far food should travel to consumers, how much farmers should be paid), a number of producers and private certification agencies have eschewed the government's labels and have begun utilizing alternative labeling terms. For example, the *Wall Street Journal* recently reported a proliferation of the alternative terms "biodynamic," "Food Alliance Certified," "beyond organic," "local," "sustainable," and "tairwa" (derived from the French word *terroir*, which translates loosely to "the essence of the land") (McLaughlin 2005).

This dissatisfaction with government-sponsored organic labels was predicted by commentator Benjamin Gutman (1999, p. 2359), who argued that because "the organic philosophy constitutes a 'continuum of attitudes a practices' rather than a concrete platform susceptible to absolute definition," any attempt at government standardization of the organic label would sacrifice ethical pluralism, stifle innovation, impair consumer decision making, and ultimately leave consumers dissatisfied. Gutman argued that divergent consumer preferences could best be honored via privately driven certification, under which private agencies would develop their own standards and would compete for consumer loyalty.

## The Kosher Approach: Government Permits, but Does Not Facilitate, Voluntary Labeling

The regime Gutman (1999) described is the second, and most laissez faire, voluntary labeling approach. Under it, the government neither facilitates voluntary labeling nor erects barriers to such labeling. Instead, the government's role is limited to policing the fraudulent use of labels.

This is the voluntary labeling regime applicable to kosher foods. Kosher status is analogous to organic status in several respects. Both depend on production and processing standards rather than product standards, and both are appealing to discrete groups of consumers, who often will pay a premium for compliant products. In addition, the terms "kosher" and "organic" may both mean different things to different people. Just as the organic community differs on matters such as the permissibility of natural pesticides and fertilizers or the wages required for farm worker, groups of observant Jews disagree on many details of the food laws—for example, whether gelatin may be deemed kosher absent a showing that it was derived from properly slaughtered, nonforbidden animals. In light of this pluralism, a single uniform kosher standard would leave many consumers dissatisfied. Groups within the religious Jewish community have therefore sought to accommodate their members' pluralistic views of the food laws by privately developing competing labels that signify a food product's kosher status. Indeed, private kosher certification agencies currently utilize more than 200 registered kosher symbols in the United States, enabling consumers to fulfill their process preferences by purchasing foods certified by the organizations they trust (Gutman 1999). A voluntary labeling system has thus developed—and thrives—without any government intervention beyond prosecution of intellectual property and fraud violations.

Between the two voluntary labeling schemes discussed thus far, it is not immediately obvious which would better inform consumers of a food's bioengineered status. On the one hand, the kosher example suggests that government standard setting may not be necessary, and the recent dissatisfaction with the government's organic standards suggests that wholly private ordering may be more desirable. On the other hand, there is likely to be less diversity of opinion over what constitutes a "nonbioengineered" or "GM-free" food product than there is over what constitutes a "kosher" or "organic" food product, so government standardization of negative labels regarding bioengineering status would likely create fewer difficulties than standardization of organic labels. In addition, private certification of kosher status succeeds in large part because the tight-knit kosher community is likely to sanction certifiers that make mistakes; that may not be true with respect to certifiers of nonbioengineered foods. In any event, either of the two voluntary labeling schemes discussed so far would likely provide a better approach to the GM issue than the third approach.

## The rBST Approach: Government Permits, but Discourages, Voluntary Labeling

Under the final voluntary labeling scheme, the government neither facilitates nor is neutral toward private labeling efforts. Instead, it technically permits, but practically discourages, voluntary negative labeling. As explained further below, this is the stance U.S. law currently takes toward the voluntary labeling of non-GM food products and is, we believe, inappropriately restrictive of sellers' efforts to respond to consumers' demands for information about the process by which a food is produced. It is also the approach American law has taken toward the voluntary labeling of milk produced from cows that have not been treated with rBST, a genetically engineered hormone known to increase weight, appetite, and milk production in cattle.

Despite scientific evidence that milk from cows treated with rBST is indistinguishable from milk produced by cows not treated with the hormone, many consumers have expressed a preference for milk from untreated cows. Accordingly, some milk producers have sought to label their milk as being from "rBST-free" cows. While such labels are literally true and would seem to accommodate heterogeneous consumer preferences, the FDA has discouraged the labels by indicating that they may subject sellers to liability unless accompanied by additional information. Specifically, the FDA presumes that such labels convey the false impression that milk from rBST-treated cows is inferior to that produced by untreated cows unless the labels are accompanied by a statement such as "No significant difference has been shown between milk derived from rBST-treated and non-rBST-treated cows" (FDA 1994a). In other words, sellers are dissuaded from engaging in negative labeling unless they also provide information that the trait they are touting, and that consumers appear to value, is ultimately irrelevant. Not surprisingly, few producers have chosen to provide the negative labels that would inform consumers as to which milk is, and is not, produced from rBST-treated cows.

There are numerous possible reasons for the FDA's stance on rBST-free (and, as discussed below, GM-free) labeling. The most benign explanation is that the agency desires to prevent consumers from being duped into believing that there is a relevant difference in the end product created by the different production processes. Alternatively, the agency may be motivated by a desire to quell irrational fears regarding innovative production processes that "tamper with nature." By discouraging sellers from playing on those irrational fears, regulators might be able to prevent the fears from spreading and becoming reinforced by social forces, as irrational fears tend to do. The least benign rationale for the FDA's approach is that the agency has been captured by large-scale agricultural interests. As explained below, there is some evidence in support of this theory.

Regardless of the FDA's reason for actively discouraging negative labeling regarding rBST, that is the tack it has taken. As described below, the agency has erected similar barriers to voluntary negative labeling of food products that are not bioengineered and do not contain GM ingredients. We next examine those barriers and argue that they are illegitimate.

## IMPEDIMENTS TO VOLUNTARY LABELING

As explained above, consumers' heterogeneous preferences regarding the use of GM foods can be accommodated through the voluntary labeling of non-GM foods by sellers who seek to satisfy those preferences. However, this conclusion is premised on a regulatory scheme that freely permits or even encourages the voluntary provision of information about the bioengineered status of non-GM foods. The current FDA regulations fail this test. Instead, they create unjustified barriers that impede the provision of such information.

While the FDA maintains that bioengineered foods are substantially equivalent to their non-GM counterparts and pose no particular health or safety risks, agency focus groups convened in 2000 revealed that many consumers are interested in whether foods contain bioengineered products. As a result, the agency conceded that some manufacturers may want to respond to this consumer desire and promulgated its *Draft Guidance for Industry: Voluntary Labeling Indicating Whether Foods Have or Have Not Been Developed Using Bioengineering* (FDA 2001). Those guidelines pose substantial barriers to the dissemination of information about food products not containing GM ingredients because they ban the terms most commonly used to describe non-GM foods.

Indeed, writing a label that will satisfy the FDA is a little like playing the popular board game Taboo, in which players provide clues to get their partner to identify a particular word but, in so doing, are forbidden to say any of the words one would most naturally use in conveying those clues. Specifically, the FDA has made the following terms "taboo":

- Acronyms such as "GM" and "GMO"
- The term "genetically modified"
- References to "organisms" or "GMOs"
- Claims to be GMO "free"
- Any implication of superiority

By making it difficult for sellers of such products to market their goods, the guidance is likely to chill the development of markets for such products, thereby depriving consumers of choice.

## Acronyms

The FDA disfavors the acronyms "GM" and "GMO" (genetically modified organism) because, it says, consumers do not understand them. Commercial experience suggests otherwise. If the FDA were correct, food producers desiring to inform their customers about the absence of bioengineered ingredients in their products would use other terms. Yet, a number of sellers have sought to label their products as "GM free" or "GMO free," apparently because they believe those acronyms catch the attention of consumers who care about GM contents. Indeed, across the world the acronyms "GM," "GMO," and "GE" (genetically engineered) are used interchangeably to refer to crops and animals that have been modified using the techniques of gene splicing or recombinant DNA. Because they convey important information to consumers who care about the issue, these acronyms should be permitted unless the FDA has evidence to indicate that they actually cause buyers to make erroneous assumptions about the contents of labeled products. The FDA has not done so.

### "Genetic Modification"

Producers will fare no better if they spell out the terms signified by the acronyms "GM" and "GMO." The FDA contends that labels denying the presence of "genetically modified" ingredients are misleading because virtually all foods have been "genetically modified" through cross-breeding (FDA 2001). Under this broad definition of genetic engineering, "[m]ost, if not all, cultivated crops have been genetically modified" (FDA 2001). Thus, a disclaimer of GM contents will usually be false. The FDA also concluded that consumers would not uncover the deception because they "do not have a good understanding that essentially all food crops have been genetically modified and bioengineering technology is only one of a number of technologies used to genetically modify crops" (FDA 2001).

Although the FDA's reasoning has a surface plausibility, that plausibility disappears upon analysis. The term "genetically modified," widely employed in international diplomacy, is a term of art in contemporary policy debates. In common parlance, GM crops are those whose genomes have been engineered using the techniques of recombinant DNA (Beales 2000). Only the U.S. government and U.S. industry assiduously avoid this terminology, preferring instead the term "bioengineered." Yet, that term would technically warrant a similarly broad definition. Strictly speaking, modern hybridization techniques could certainly be characterized as bioengineering: They, too, involve the biological manipulation of crop and livestock genomes in order to produce desired attributes. By convention, however, the term "bioengineering," like the term "genetic modification," is used to describe the alteration of an organism's DNA using the techniques of recombinant DNA.

Between the two terms, genetic modification is far more commonly used. Bookseller Amazon.com, for example, lists 143 books with "GMO" or "genetically modified organism" in the title and only three titles using some form of the word "bioengineer." Furthermore, the phrase "genetic modification" seems more informative because it tells the consumer that existing genes have been modified, a fact that is not made explicit by the term "bioengineered." Thus, the FDA's strong preference for the term "bioengineering" cannot be explained by concerns about linguistic clarity.

A more likely explanation for the FDA's desire to eliminate the term "genetic modification" from domestic discourse is the common use of that term by critics of biotechnology. Those critics normally describe the plants produced by gene splicing as "genetically modified." By contrast, the FDA had previously learned in focus group sessions that the "bio" prefix invoked positive connotations. It hardly seems coincidental that the FDA's ensuing guidance attempts to substitute the positive term "bioengineering" for references to "genetic modification."

### The Absence of "Organisms"

The FDA also claims that a food label touting the absence of GMOs is misleading because it implies that foods which are not GMO-free contain "organisms"—that is, living things. This claim is misleading in the eyes of the FDA "because most foods do not contain organisms" (FDA 2001). If the FDA is suggesting that buyers of GMO-free products will assume that a food labeled as lacking GMOs contains fewer microorganisms, such as bacteria, than foods that are not negatively labeled, it is disingenuously ignoring the common language of the debate over GM foods. Genetically modified corn and soybean plants, like the GM bacteria that scientifically preceded them, are genetically modified living organisms. The corn and soybeans they produce are routinely described as "GMOs," and food products that utilize such products are said to "contain" GMOs. By contrast, foods that do not contain such products are commonly understood to be "GMO free." In the online Wikipedia encyclopedia, for example, GM food is defined as "a food product containing some quantity of any genetically modified organism (GMO) as an ingredient" (Wikipedia 2005).

## GM "Free"

The FDA also objects to labels claiming that the contents of the package are "free" of bioengineered ingredients. The FDA plausibly contends that a claim that a product is GM "free" implies that zero GM material is present. Because food producers will rarely be able to guarantee the absence of trace elements of GM materials, the FDA has directed sellers to avoid the word "free" altogether.

The FDA's argument about trace contaminants certainly seems logical. It is also consistent with the comments the FDA received in its 2000 focus groups, in which participants expressed a belief that "free" means zero (FDA 2000). Yet, the authors of the guidance certainly knew that their insistence on purity was inconsistent with prior practice at the FDA. The agency long ago abandoned a zero-tolerance policy for other food ingredients. Under the agency's food labeling guidelines, food products can be labeled as "free" from sugar, fat, sodium, cholesterol, and other substances as long as the trace contamination does not exceed tolerances set by the FDA (FDA 1994b). For example, the FDA food labeling guidelines permit up to 0.5 grams per serving of fat in a "fat-free" product. The USDA's new organic food labeling rules take a similar approach, permitting up to 5% nonorganic ingredients in a food labeled "organic" (USDA 2006). Although the FDA claims that it lacks the testing technology needed to permit a similar threshold for GM-free foods, that claim is not convincing.

There is an additional reason for allowing trace elements of GM crops to appear in GM-free foods. Despite repeated requests to do so, neither the U.S. government nor the states have enacted laws to prevent cross-fertilization or to require identity preservation (Endres 2005). Organic and non-GM farmers have long complained that this regulatory vacuum guarantees the contamination of their products. Given this regulatory inertia, it seems unfair for regulatory policy to insist on absolute purity. Doing so would allow Monsanto and the other GM seed producers to eradicate their competition by contaminating their products.

The FDA does offer an alternative means of asserting that GM ingredients have not been used. It suggests that food sellers provide a more detailed description of the process that was used to process the food. For example, a seller's label may say "We do not use ingredients that were produced using biotechnology" (FDA 2001). Yet, this claim also seems to assume purity. Perhaps its acceptability stems from the fact that it is much tougher to catch a consumer's eye with statements like this than with a bold "GM-free" label. That, in the end, appears to be the point of these rules.

## Claims of "Superiority"

The broadest of the FDA's restrictions is its prohibition of any implication that non-GM foods are superior (FDA 2001). Since the only purpose for indicating that a food has no GM ingredients is to appeal to shoppers who believe, for one reason or another, that non-GM foods are superior, this rule effectively prohibits any mention of the absence of bioengineered ingredients unless the label also contains a detailed disclaimer indicating that the absence of GM ingredients is actually irrelevant. To enforce this policy, the FDA sent letters to several organic food companies warning them not to label their products "GMO free"; doing so, the agency contended, suggested that GM products are inferior (Pew Initiative on Food and Biotechnology 2002). As a result, Spectrum Organic Products, Inc., a company that goes as far as France to get oil guaranteed to be untainted by biotechnology, reduced its mention of non-GM contents to small print on the back of the bottle.

Perhaps the imposition of these barriers to voluntary labeling would be justified if there were no rational basis upon which a consumer could believe that non-GM foods are superior. In that event, the agency could conclude that "GM-free" labels are intended to imply and to capitalize upon the false belief that GM ingredients are unsafe. Yet, there are several quite rational reasons why some people prefer to avoid GM foods. Some people believe that non-GM foods are superior because they dislike some aspect of the process that produced them. Others are not confident relying on the FDA's assurances about their safety, want long-term data not yet available, or simply have a high level of aversion to novel risks. None of these reasons for preferring GM-free foods is irrational, and even if only a small fraction of Americans currently prefer non-GM foods (a fact yet to be ascertained), food producers should be permitted to respond to their preferences with the same zeal as the seller of any other product.

### Nonsafety Concerns

Because the FDA's function is to preserve food safety, it seems to have assumed that all claims that GM-free foods are superior imply that GM-free foods are safer, a contention that the FDA vigorously rejects. Yet, critics of GM foods have expressed a variety of nonsafety objections to them. Some of these critics feel, for example, that the purchase of GM foods would implicate them in the environmental risks that they associate with some GM crops, such as corn, and animals, such as salmon. Other objections stem from the concentration of economic and social power residing in the high-tech seed manufacturers. Critics also fear that reliance on these seeds will extend the dominance of industrialized agriculture, a phenomenon that already threatens the survival of family farms and rural towns. Environmentalists often express a fundamental moral objection to gene splicing that works in ways that nature could not, such as transferring genes across the species boundary. They also fear that a shift to high-tech, patented seeds will deprive farmers of their independence and make farmers in developing countries dependent on the political and economic power of supernational corporations. Given the importance of these ethical, social, cultural, political, and economic issues to some consumers, sellers ought to be encouraged to inform consumers that non-GM products are available. This will enable consumers to "vote" on these policy issues through their purchasing decisions (Kysar 2004).

Many of these concerns arise out of the process of gene splicing and, thus, would persist even if the safety of GM foods were incontrovertible. In this respect, consumers of non-GM foods are hardly unique. *Processes* are often important to consumers even when the process does not change the quality, safety, or utility of the resulting *product* (Kysar 2004). That is why products are labeled "union made," "fair trade," "dolphin-safe," and "made in the USA." Concerns about the process of production also provide an important part of the reason for labeling foods "organic."

## Safety Concerns

Even on the issue of safety, there may be rational reasons for preferring nonbioengineered foods. Consumers understand that the initial safety assessments of any new technology can never entirely rule out risk, especially those risks that only express themselves with long-term use. Vioxx, DDT, and fen-phen have reminded them of this reality.

Participants in the FDA's 2000 focus groups emphasized the uncertainty about long-term risks much more than the social and environmental risks (FDA 2000). They acknowledged that scientists had largely ruled out any serious short-term risks from the consumption of the GM crops currently on the market. Yet, they were so concerned about the uncertainties associated with long-term use that they favored *mandatory* labeling almost unanimously. They told the FDA that prior technological innovations, such as pesticides, had been introduced mainly for the sake of producers, rather than consumers, and yet they perceived that consumers had suffered the long-term harms. Some focus group participants feared that consumers of GM foods were acting as "guinea pigs"; others doubted the ability of the FDA to withstand pressure from industry. Labeling, they felt, would enable them to tell it was a new technology.

Experts agree that gene splicing does pose its own unique risks. For that reason, the National Academy of Sciences has concluded that the risks of an unforeseen, harmful mutation are more likely using the techniques of recombinant DNA than with traditional forms of hybridization (Institute of Medicine and National Research Council 2004).

Some consumers prefer not to expose themselves to these risks given the minor benefits that they feel they receive from access to GM products. These consumers prefer to "wait and see." For them, GM-free foods will often be superior. Their preference for caution in the face of uncertainty, while not shared by the FDA, is certainly a legitimate and "rational" basis for their choice.

Even if the FDA correctly concluded that "concern about the unknown" does not provide a sufficient basis for *mandating* disclosure (FDA 2001), the absence of information about the long-term risks of GM foods certainly provides a legitimate basis for consumers to prefer GM-free foods and, thus, a basis for voluntary labeling. Sellers can satisfy these consumer preferences only if they can clearly and prominently label their products as GM free. Yet, the FDA seems determined to squelch this kind of marketing.

## RATIONALES FOR RESTRICTING VOLUNTARY LABELING

Why would the FDA exhibit this sort of bias against sellers who want to market their products as GM free? Several answers are plausible. Most benignly, the FDA may fear that bold and simple labeling will fuel irrational fears of biotechnology. Alternatively, the FDA may feel the need to conform its policies to the positions that the U.S. government has taken in the international disputes involving GM crops. In those disputes, the United States has argued that the mandatory labeling rules imposed by many other countries are a form of illegal protectionism. Finally, the FDA may be catering to its industry clients.

### Discouraging Irrational Consumer Preferences

The FDA appears to fear that the prominent labeling of GM-free foods will suggest to consumers that GM foods are unsafe. Because scientists have found no credible evidence of additional health risks, at least with respect to short-term use, the FDA has concluded that any label creating an inference of superior safety is misleading. Although it reached this conclusion before a persuasive body of evidence existed, the agency's educated hunch has now been confirmed. Today, the FDA is so concerned about consumer misconceptions about the health risks associated with GM foods that it requires disclaimers about GM safety even when the seller has stated only that the product is "GM free."

Is the agency's fear of a boycott by misinformed consumers justified? Studies show that the public often perceives a risk to be higher than experts do (Breyer 1993; Sunstein 2002). A number of mental and environmental factors contribute to these errors. As described above, individuals are subject to heuristics that can cause them to amplify the risks associated with unnatural occurrences (i.e., processes that involve "tampering with nature") and to give greater credit to more alarmist accounts of circumstances. The resulting alarm is often reinforced by news media inclined toward sensationalism and by social forces that lead individuals to accept propositions simply because they believe others accept them.

The public suffers when it falls prey to misconceptions about risk. As a matter of social utility, public mistakes about product riskiness could lead confused consumers to make choices that are not preference maximizing. From a rights-based perspective, consumer confusion could decrease consumer autonomy. Says F. Scott Kief, a law professor at Washington University in St. Louis, "If they're buying it because they think it's healthier, and it turns out not to be, then they are being harmed" (Bess 2003).

Psychologists have found evidence that ordinary people tend to have extreme reactions to risk. We tend toward one of two responses: "alarmist overreaction (take precautions, better-safe-than-sorry), or complete neglect (out-of-sight, out-of-mind)" (Henderson and Rachlinski 2000: 255). Below a certain level of objective significance, consumers fail to perceive a risk at all. Beyond that certain level, however, consumers overestimate the risk. Moreover, Henderson and Rachlinski (2000, p. 254) suggest that overestimation is easier to prompt than underestimation: "The human brain seems, on the whole, built to overreact to risk rather than to casually disregard risk."

Supporters of biotech foods contend that the reaction of European consumers to GM foods is a perfect example of overreaction to unsubstantiated fears. In the United Kingdom, a public campaign by Greenpeace effectively forced GM products off the shelves of all major grocers.

Thus far, consumers in the United States have not shown similar interest in the process used to produce their foods. However, studies in the United States have shown significant public distrust of GM foods and widespread demand for mandatory labeling. Furthermore, the industry worries that "fear mongers" such as Greenpeace, the Sierra Club, the Union of Concerned Scientists, and the Consumer Federation of America will attempt to take advantage of this skepticism. Because the FDA considers this public skepticism to be unfounded, its strict rules on voluntary labeling may be intended to protect GM seed producers and the farmers who use their seeds from harm caused by unfounded consumer fears.

Nevertheless, there are good reasons to believe that this fear of public irrationality is overstated and that the harm caused by the FDA's restricting labeling rules outweighs its purported benefits. First, on balance, people are more prone to underestimate the risks of everyday activities over which they exert some control than to overestimate them. Many studies have found that people across demographic categories are optimistic with respect to the personal risks posed by a seemingly endless array of products and activities (Hanson and Kysar 1999). Smokers are a prime example. Despite a concerted public health campaign to educate the public about the risks, people routinely underestimate the risks to themselves.

Second, manufacturers are quite capable of responding to unsubstantiated claims that biotech foods are unsafe. In this country, they have been doing so successfully for more than a decade—too successfully, in the view of those critics who are concerned about unknown long-term risks. Under the current status quo, the government effectively *subsidizes* sellers of GM foods by preventing them from having to expend resources to educate consumers.

Third, the lesson of the FDA's 2000 focus groups is that consumers are unlikely to misread voluntary food labels. The FDA conducted 12 focus groups across the country between May 10 and 24, 2000, as part of its preparations to draft the guidance on labeling (FDA 2000). In these sessions, the FDA learned that American consumers are not biased against the technology. Most had heard about the technology as one with great potential. Most "accepted as a matter of course that the short-term safety of bioengineered foods can be determined by science and therefore it was not in question" (FDA 2000). In the focus groups, a strong consensus emerged that unknown long-term health consequences that cannot be anticipated based on current science or knowledge were the major concern.

The participants assured the FDA that they would not interpret a GM-free label as adding anything to their prior understandings of the relative advantages of GM foods compared to other foods. Instead, the connotations that they attached to GM contents would be based on their preexisting beliefs about food biotechnology. According to a report on the focus group sessions, "[t]he idea that such claims need a disclaimer seemed forced and unnecessary to most participants, and it was often interpreted as a partisan stance that is pro-food biotechnology" (FDA 2000).

Fourth, the current restrictions on voluntary labeling impose a considerable burden on consumers whose preferences to avoid GM foods are not based on scientific misconceptions. Long-term risks are still unknown, and some consumers may prefer to wait and see. Other consumers want labels so that they can act on their personal beliefs about the process of food production. As Douglas Kysar (Kysar, 2004) has noted, in our increasingly globalized and deregulated world, "voting" with our pocketbooks will sometimes be the best way to express our political views. By freely labeling foods that do not contain GM ingredients, sellers can empower those consumers who prefer to avoid GM foods because of concerns about the social and environmental impact of the process used to make these foods.

The FDA seems to have forgotten that process information can be very relevant to consumers for reasons completely unrelated to food safety. Although these reasons may not provide a legal basis to mandate the labeling of GM contents under existing statutes, they nevertheless provide an ample basis for voluntary labeling.

Consumers who wish to express preferences of this kind through their purchasing decisions are not simply buying a product. They are expressing their political and social beliefs. The FDA's restrictive labeling rules interfere with their ability to do so. As a consequence, the current labeling rules constitute a form of political disenfranchisement that raises important First Amendment issues.

If there is one fixed principle in the commercial speech arena, it is that "a State's paternalistic assumption that the public will use truthful, nonmisleading commercial information unwisely cannot justify a decision to suppress it" (*44 Liquormart, Inc. v. Rhode Island* 1996). In its 1996 *Liquormart* decision, the Supreme Court stated the principle quite clearly: "The First Amendment directs us to be especially skeptical of regulations that seek to keep people in the dark for what the government perceives to be their own good." Furthermore, when the label's contents are accurate, the federal government cannot prohibit the seller's disclosure on the grounds that it has the "potential" to mislead the buyer. Only speech that is "inherently misleading" can be censored (*Ibanez v. Florida Dep't of Business & Professional Reg.* 1994). As a consequence, labels that accurately indicate the absence of GM ingredients are a form of protected speech and cannot be prohibited.

Given the burdens that the FDA's current restrictions impose on consumer choice and freedom of speech, the FDA's understandable, but highly speculative, fear that consumers will be duped into boycotting GM foods because of misconceptions about product safety does not justify the current restrictions on voluntary labeling. Given the array of factors that drive the debate over GM foods, the FDA should be encouraging, rather than discouraging, voluntary labeling so that consumers can act on their preferences. Sellers of GM food have no more right to override consumer preferences than chicken producers have to override a consumer preference for "the other white meat."

## Agency Capture and U.S. Foreign Trade Policy

The degree of resistance that the FDA has shown to voluntary labeling is puzzling in some respects. A robust market in voluntary labels offers an attractive compromise between the boosters of GM food and its critics. Voluntary labeling would permit GM seed manufacturers and farmers to sell their products unless and until more concrete evidence of danger surfaces, while at the same time letting consumers who fear the long-term uncertainties or who object to GM foods on moral, environmental, or sociopolitical grounds to opt out. By encouraging voluntary labeling, the government can continue its support of the biotechnology industry while showing that it respects the views of all consumers, even those who disagree with its opinions. In addition, voluntary labeling is consistent with the current administration's professed preference for market-driven solutions.

Yet, the FDA has taken a very different path. While agency concern about irrational consumer decision making offers one possible explanation, politics provides another. It is quite possible that the FDA's actions are best explained as the product of political pressure from the White House and industry lobbyists to be a "team player" and to cooperate with long-standing agricultural and foreign trade policies that prioritize GM crops and their beneficiaries. The biotechnology industry is no more eager for competition than any other favored industry.

### Agency Capture

Because political influence often operates in the shadows, suspicions of this kind are difficult to corroborate. However, several pieces of evidence are consistent with the hypothesis of agency capture.

When Dan Glickman was the Secretary of Agriculture in the Clinton administration, he told a reporter that he was pressured by the administration and the industry to be a cheerleader for a technology that he was regulating (Lambrecht 2001, p. 139). "What I saw generically on the probiotech side," he said,

> was the attitude that the technology was good and that it was almost immoral to say it wasn't good because it was going to solve the problems of the human race and feed the hungry and clothe the naked. There was a lot of money that had been invested in this, and if you're against it you're Luddites.... That, frankly, was the side our government was on.

He went on to say that the issue had been framed as a "trade issue" and that opponents were seen as foolish or as foreigners who "wanted to keep our product out of their market" (Lambrecht 2001, p. 139).

Furthermore, the person who signed off on the FDA's first template for biotech labeling was a lawyer for Monsanto both before and after his term at the FDA. In 1994, the FDA had imposed strict restrictions on the labeling of milk from cows that had not been treated with the genetically engineered enzyme rBST (FDA 1994a). Its use was controversial in the United States among environmentalist organizations and animal rights groups and is prohibited in the European Union. Because this enzyme also occurs naturally in cow milk, the FDA challenged producers who labeled milk from untreated cows as "rBST free." The FDA felt that this label created the misimpression that milk from untreated cows was different in composition than milk from cows treated with rBST. In the milk itself, the enzymes are identical. As result, the FDA insisted that milk labels refer only to the use of rBST in the process of milk production and make no claim about differences in the milk itself. In addition, the labels had to include an affirmative disclaimer indicating the production of milk from cows treated with rBST does not affect the milk.

The FDA's 1994 guidance was written by Michael R. Taylor, an attorney who had represented Monsanto, the manufacturer of rBST, for 10 years (1981–1991) as a partner at King and Spaulding before becoming Deputy Commissioner for Policy at the FDA. He left the FDA in 1994 after authoring this guidance, returned to King and Spaulding for two years, and then became head of Monsanto's Washington, D.C. office from 1998 to 2000. Critics claim that the FDA guidance that he signed closely mirrors Monsanto's lobbying position (Kysar 2004).

Capture is also suggested by the terms of the labeling rules. They reflect an extraordinarily aggressive effort to control the language of the public debate over GM foods. Consider, for example, the FDA's insistence on absolute purity when conventionally bred crops are marketed as "GM free." Trace amounts are permitted when Big Food wants to market products as "fat free" but not when a competitor to Big Food wants to market products as "biotech free." Similarly, the FDA permits "added-value" statements, such as statements about the absence of "nonnutritive substances" such as "artificial colors" (FDA 2000), whereas added-value statements about the absence of GM ingredients require disclaimers. It appears that the FDA does not want prominent labeling of GM-free foods.

## U.S. Agricultural and Foreign Trade Policy

In both domestic and foreign policy arenas, several recent presidential administrations have also strongly endorsed the biotechnology industry contention that GM crops are identical in all material respects to the grains produced using conventional seeds. From this vantage point, the process used to produce a product is irrelevant if the end product is substantially the same. Since at least 1991, this has also been the policy of the domestic regulatory agencies, such as the FDA and the USDA. It underpins the current U.S. complaint before the World Trade Organization about European restrictions on the marketing of GM crops (Grossman 2005). In the international arena, we have vigorously advanced the position that the process used to produce these crops is completely irrelevant.

That makes it very hard to accede to the requests of domestic consumers for prominent labeling. The fact that many consumers disagree with the U.S. government's conclusion that the process of production is irrelevant, including most participants in the FDA's focus groups, could potentially be very embarrassing for the United States. Citizens who want information about GM contents present a serious dilemma for the regulatory agencies because they want information that the U.S. government strongly contends is completely irrelevant.

Given the U.S. government's vigorous endorsement of GM crops as a matter of agricultural policy and international trade policy and its strong endorsement of the product/process distinction, the FDA presumably faced considerable political pressure from inside and outside the agency to prevent an embarrassing revolt of domestic consumers against GM foods. It could help to prevent this catastrophe by imposing strict restrictions on voluntary labeling that push information about the process of production off the front of the label and into the fine print on the back. That is what it did.

## CONCLUSION

Policy makers throughout the world, even in the GM-hostile European Union, are moving away from direct restrictions on the cultivation of GM crops. Presumably, this trend reflects a growing consensus that bioengineering's potential benefits are likely to outweigh its costs (in particular, any negative externalities it creates). At the same time, however, many countries require the labeling of GM products, and critics of GM products would like mandatory labeling rules here. So far, the debate over GM labeling proposals has been rather polarized, with one camp advocating mandatory labeling and the other defending the American status quo as reflected in the FDA's voluntary labeling guidance.

Our goal in this chapter is to propose a middle ground. For the reasons detailed herein, we do not believe that mandatory labeling represents, at least at this point in time, the optimal labeling scheme. Instead, consumer information should be provided via a voluntary negative labeling scheme, which will present consumers with autonomy-enhancing information at a lower cost. The American status quo, however, actively *discourages* the sort of voluntary labeling that would most efficiently satisfy consumer demands for information regarding a food product's bioengineered status.

Our "middle ground" position is that regulators should remove the barriers to consumer information regarding GM content, thereby permitting voluntary negative labeling to flourish and eliminating the need for a mandatory positive labeling scheme. In particular, the FDA should abandon its presumption that true claims about a lack of bioengineered content are misleading. Many consumers want this information and sellers should be encouraged to provide it. Many consumers believe that the process used to produce the crop *is* relevant, even if the resulting product is substantially the same. Some also question the U.S. government's claim that the foods produced by gene splicing pose no additional health risks, especially for long-term use. Sellers should be permitted satisfy their preferences. Under current rules, however, these sellers cannot aggressively market their products as GM free. Instead, truthful and relevant information is forced into the "footnotes." Given the current administration's preference for market solutions in other contexts, the continuation of these rules strongly suggests FDA capture by either the biotechnology industry or its friends in U.S. foreign trade policy.

## References

*44 Liquormart, Inc. v. Rhode Island,* 517 U.S. 484 (1996).

Baker, Gregory, and Burnham, Thomas. 2001. Consumer Response to Genetically Modified Foods: Market Segment Analysis and Implications for Producers and Policy Makers. *J. Agric. & Resource Econ.* 26:387.

Beales, J. Howard. 2000. Modification and Consumer Information: Modern Biotechnology and the Regulation of Information. *Food & Drug L. J.* 55:105.

Bernauer, T., and Meins, E. 2001. *Scientific Revolution Meets Policy and the Market: Explaining Cross-National Differences in Agricultural Biotechnology Regulation.* Discussion Paper No. 0144. Adelaide, Australia: University of Adelaide Center for International Economic Studies.

Bess, Allyce. Aug. 10, 2003. Labeling Dairy Products: Got Posilac? *St. Louis Post-Dispatch.*

Breyer, Stephen. 1993. *Breaking the Vicious Circle: Toward Effective Risk Regulation.* Cambridge, Mass.: Harvard University Press.

Carter, Colin A., and Gruère, Guillaume P. 2003. Mandatory Labeling of Genetically Modified Foods: Does It Really Provide Consumer Choice? *AgBioForum* 6:68.

Cauvin, Henri E. Sept. 4, 2002. Zambian Leader Defends Ban on Genetically Altered Foods. *New York Times,* A6.

Chern, Wen S., et al. 2002. Consumer Acceptance and Willingness to Pay for Genetically Modified Vegetable Oil and Salmon: A Multiple-Country Assessment. *AgBioForum* 5:105.

Endres, A. Bryan. 2005. Revising Seed Purity Laws to Account for the Adventitious Presence of Genetically Modified Varieties: A First Step toward Coexistence. *J. Food Law & Pol'y* 1:131.

FDA (Food and Drug Administration). Feb. 10, 1994a. *Interim Guidance on the Voluntary Labeling of Milk and Milk Products from Cows That Have Not Been Treated with Recombinant Bovine Somatotropin.* Fed. Reg. 59:6279.

FDA (Food and Drug Administration). 1994b. *A Food Labeling Guide—Appendix A, Definitions of Nutrient Content Claims,* vm.csfsan.fda.gov/~dms/flg-6a.html.

FDA (Food and Drug Administration). 2000. *Report on Consumer Focus Groups in Biotechnology,* www.cfsan.fda.gov/~comm/biorpt.html.

FDA (Food and Drug Administration). 2001. *Draft Guidance for Industry: Voluntary Labeling Indicating Whether Foods Have or Have Not Been Developed Using Bioengineering,* www.cfsan.fda.gov/~dms/biolabgu.html.

Fisher, Cynthia. 2002. Note, The Genie Is Out of the Bottle: Consumers Demand Mandatory Labeling on Genetically Engineered Foods. *J. Legal Advoc. & Prac.* 4:88.

Grossman, Margaret Rosso. 2005. Traceability and Labeling of Genetically Modified Crops, Food, and Feed in the European Union. *J. Food & Pol'y* 1:43.

Gutman, Benjamin. 1999. Note, Ethical Eating: Applying the Kosher Food Regulatory Regime to Organic Food. *Yale L. J.* 108:2351.

Hadfield, Gillian K., and Thomson, David. 1998. An Information Based Approach to Labeling Biotechnology Consumer Products. *J. Consumer Pol'y* 21.

Hanson, Jon, and Kysar, Douglas. 1999. Taking Behavioralism Seriously: The Problem of Market Manipulation. *N.Y.U. L. Rev.* 74:630.

Hanson, Jon, and Kysar, Douglas. 2000. Taking Behavioralism Seriously: A Response to Market Manipulation. *Roger Williams U. L. Rev.* 6:59.

Henderson, James, and Rachlinski, Jeffrey. 2000. Product-Related Risk and Cognitive Biases: The Shortcomings of Enterprise Liability. *Roger Williams U. L. Rev.* 6:213.

*Ibanez v. Florida Dep't of Business & Professional Reg.*, 512 U.S. 136 (1994).

Institute of Food Technologists. 2000. IFT Expert Report on Biotechnology and Foods: Labeling of rDNA Biotechnology-Derived Foods. *Food Technology* 54:63.

Institute of Medicine & National Research Council. 2004. *Safety of Genetically Engineered Foods: Approaches to Assessing Unintended Health Effects.* Washington, D.C.: National Academy Press.

Kalaitzandonakes, N., and Bijman, J. 2003. Who Is Driving Biotechnology Acceptance? *Nature Biotechnology* 21:366.

Kuran, Timur, and Sunstein, Cass. 1999. Availability Cascades and Risk Regulation. *Stan. L. Rev.* 51:683.

Kysar, Douglas. 2004. Preferences for Processes: The Product/Process Decision and the Regulation of Consumer Choice. *Harv. L. Rev.* 118:525.

Lambert, Thomas. 2004. Avoiding Regulatory Mismatch in the Workplace: An Informational Approach to Workplace Safety Regulation. *Neb. L. Rev.* 82:1006.

Lambrecht, Bill. 2001. *Dinner at the New Gene Café.* New York: Thomas Dunne Books.

Mandel, Gregory. 2004. Gaps, Inexperience, Inconsistencies, and Overlaps: Crisis in the Regulation of Genetically Modified Plants and Animals. *Wm. & Mary L. Rev.* 45:2167.

Marden, Emily. 2003. Risk and Regulation: U.S. Regulatory Policy on Genetically Modified Food and Agriculture. *B.C. L. Rev.* 44:733.

McCluskey, Jill, et al. Sept. 21, 2001. *Consumer Response to Genetically Modified Food Products in Japan.* Research Paper TWP-2001-101, Washington State University, impact.wsu.edu/research/twp/01-101.pdf.

McLaughlin, Katy. Apr. 19, 2005. Is Your Tofu Biodynamic? Making Sense of the Latest Organic Food Terminology. *Wall St. J.*, D1.

Mendenhall, Catherine, and Evenson, Robert. 2002. Estimates of Willingness to Pay a Premium for Non-GM Foods: A Survey. In *Market Development for Genetically Modified Foods*, ed. Vittorio Santaniello et al. New York: CABI Publishing, 55.

Miller, Henry. 2003. Vox Populi and Public Policy: Why Should We Care? *Nature Biotech.* 21:1431.

National Organic Program, 7 CFR Part 205 (2002).

Noussair, Charles, et al. 2004. Do Consumers Really Refuse to Buy Genetically Modified Food? *Econ. J.* 114:102.

Pew Initiative on Food and Biotechnology. 2001. *Public Sentiment about Genetically Modified Food*, pewagbiotech.org/research/gmfood/survey3-01.pdf.

Pew Initiative on Food and Biotechnology. Nov. 1, 2002. News in Brief. *AgBiotech Buzz* 2(10), pewagbiotech.org/buzz/display.php3?StoryID=85.

Pew Initiative on Food and Biotechnology. 2004. *Overview of Findings: 2004 Focus Groups and Polls*, pewagbiotech.org/research/2004update/overview.pdf.

Preston, Christopher. 2005. *Peer Reviewed Publications on the Safety of GM Foods*, www.agbioworld.org/biotech-info/articles/biotech-art/peer-reviewed-pubs.html.

Riddick, Thomas, Reavey, William, and Michels, Dirk. 1997. Private Legal Mechanisms for Regulating the Risks of Genetically Modified Organisms: An Alternative Path within the Biosafety Protocol. *Envtl. L.* 4:1.

Rousu, Matthew, et al. 2004. Are United States Consumers Tolerant of Genetically Modified Foods? *Rev. Agric. Econ.* 26:19.

Rozin, Paul. 2001. Technological Stigma: Some Perspectives from the Study of Contagion. In *Risk, Media, and Stigma*, ed. James Flynn et al. London: Earthscan, 31.

Stigler, George. 1961. The Economics of Information. *J. Pol. Econ.* 69:213.

Streiffer, Robert, and Rubel, Alan. 2003. Choice versus Autonomy in the GM Food Labeling Debate. *AgBioForum* 6:141.

Streiffer, Robert, and Rubel, Alan. 2004. Democratic Principles and Mandatory Labeling of Genetically Engineered Foods. *Pub. Affairs Q.* 18:223.

Sunstein, Cass. 1988. Beyond the Republican Revival. *Yale L. J.* 97:1539.

Sunstein, Cass. 2002. The Laws of Fear. Review of Paul Slovic, *The Perception of Risk* (2000). *Harv. L. Rev.* 115:1119.

Sunstein, Cass. 2004. Moral Heuristics and Moral Framing. *Minn. L. Rev.* 88:1556.

Tegene, Abebayehu, et al. 2003. The Effects of Information on Consumer Demand for Biotech Foods: Evidence from Experimental Auctions. *USDA Tech. Bull.* 1903:24.

Teisl, Mario, et al. 2003. Labeling Genetically Modified Foods: How Do US Consumers Want to See It Done? *AgBioForum* 6:48.

Thue-Vasquez, Diane. 2000. Genetic Engineering and Food Labeling: A Continuing Controversy. *San Joaquin L. Rev.* 10:77.

USDA (U.S. Department of Agriculture). 2006. National Organic Program Regulations, 7 e-CFR Sec. 205.302(b), www.ams.usda.gov/nop/standards/FullRegTextOnly.html, accessed Oct. 23, 2006.

Viscusi, W. Kip. 1997. Alarmist Decisions with Divergent Risk Information. *Econ. J.* 107:1657.

Wikipedia. 2005. Genetically modified food, www.reference.com/browse/wiki/Genetically_modified_food, accessed Oct. 6, 2005.

# 10

# Labeling Genetically Engineered Foods

## Rights, Risks, Interests, and Institutional Options

*Clark Wolf*

In 2002, a ballot initiative was introduced in Oregon that would have required that foods containing genetically engineered (GE) components bear a label. Oregon's Measure 27 was both broad and specific: It would have required labeling for "all foods derived in whole or in part from any genetically engineered microorganism, plants or livestock, if that genetically engineered material accounts for more than one tenth of one percent of the weight of the product," as well as for all foods prepared with the use of GE enzymes (whether or not these enzymes were present in the final product), all foods derived from GE inputs, and all meat and dairy products from livestock that had been fed feed with GE-derived ingredients. The measure further required that very specific information be included on the labels themselves. It required that "foods resulting from trans-species gene transfers" must specify on the label the "source of the transgene used, and the purpose of the transfer. For instance, 'This squash contains viral genetic information designed to make it resistant to viral infection.'" Where the process of genetic engineering involved the importation of animal genes into plants, the Oregon initiative would have required that the source be clearly identified to allow "vegetarians and those with dietary religious restrictions [to] observe their dietary guidelines. For instance, 'this tomato contains genetic material derived from the flounder, a fish of the family Bothidiae.'"[1]

Oregon's Measure 27 initially received enthusiastic support from some Oregonians. It also caught the attention of the FDA and the biotech industry. In October 2002, Lester Crawford, then Deputy Commissioner for the FDA and former director of Georgetown University's Center for Food and Nutritional Policy, sent a letter to Oregon Governor Kitzhaber, noting FDA objections to the measure. Crawford's (2002) letter emphasized the FDA's scientific judgment that "there is no significant difference between foods produced using bioengineering, as a class, and their conventional counterparts." He also noted that the Oregon labeling initiative would not conform to the FDA's traditional standards for labeling and suggested that "the proposed legislation would impermissibly interfere with manufacturers' ability to market their products on a nationwide basis" and that it would impede "the free flow of commerce between the states." Biotechnology corporations weighed in with the power of the purse. By some estimates, biotech corporations spent $5.3 million in their effort to defeat this initiative, compared with the $200,000 spent by advocates (Donohoe 2003). When Measure 27 was defeated, with 73% of Oregonians voting against its implementation, some charged that the outcome had been purchased with corporate cash. Others breathed a sigh of relief.

Should we be relieved or disturbed by the failure of Measure 27? The question is not simply an academic one. Other relevantly similar legislative measures are currently under consideration, such as the 2005 Genetically Engineered Food Right to Know Act, sponsored by California Representative Barbara Boxer and Ohio Representative Dennis Kucinich.[2] This chapter examines some of the considerations that are relevant for the evaluation of proposals to institute labels for foods containing (or produced using) GE agricultural products.

## LABELING GE FOODS: THE POLITICAL PREDICAMENT AND THE LIKELY EFFECTS OF LABELING

It is valuable to keep clearly in mind what the likely consequences of a labeling regime would be: What effect would different labeling regimes have on consumer demand? Whose interests would be advanced, and whose set back, by different labeling regimes that might be implemented? Once these have been considered, we then examine whether any of those whose interests would be set back have a right or a valid claim against those who might implement such a regime that should effectively defeat such a proposal. This way of proceeding allows us to keep separate the different kinds of morally relevant issues that come to play in this context. In considering alternative proposals, it is also valuable to identify different institutional options we might implement and to consider separately the welfare and rights implications of each one.

I note from the start that I am not persuaded by those who urge that the consequences of implementing a labeling regime would be terrible. Given evidence concerning consumer preference for alternatives that do not contain GE ingredients, one might reasonably predict that such a regime would lead to slightly lower prices for some food commodities—those bearing labels indicating that they contain GE materials—with a concomitant but larger rise in the price and availability of food products that are "GE free." Given the relative size of the organic and conventional food markets, the advantages would likely be focused on the relatively smaller number of organic producers, while the disadvantages would probably be minimal and spread out among a large number of conventional producers. Thus, we might expect to see a significant benefit for organic and non-GE producers and a slight cost, in terms of forgone profits, for producers who grow and market GE food crops.

If the effect on conventional production were more severe than this, one might expect that it could set back research in plant and animal biotechnology, with associated opportunity costs that are difficult to measure. While some people believe that biotechnology holds the richest opportunities for the future, others regard it with distaste and horror. Some are afraid that our venture into biotechnology and genetic engineering will lead to disaster. Obviously, people with these different views of biotechnology will have a different view of the opportunity costs associated with a setback in biotechnology research. Since different views of this question are deeply informed by the fears and hopes people bring to the issue, it is only fair to place my own potential biases on the table: My own view of these costs is informed by my hope that biotechnology may improve agriculture and perhaps move existing agricultural institutions closer to a sustainable system. For this reason, I judge that the opportunity costs may be significant and that a setback for biotechnology research would be a matter for concern. But this concern is mitigated by consideration of the present direction of research in plant biotechnology, which has in many ways fallen short of its promise to improve the unsustainable agricultural practices that have become traditional in the American Midwest and many other parts of the world and are rapidly spreading elsewhere. In the absence of appropriate incentives, biotechnology research can be expected to follow market forces, and the existing market forces have rewarded large farms with intensive inputs. The environmental costs of contemporary agricultural practices are significant, and these costs are already coming due. As in so many other areas, we are engaged in long-term deficit spending for which costs will fall on subsequent generations. Of course, these problems cannot be entirely attributed to biotechnology, but thus far, biotechnology has not delivered on its promise to provide us with a more sustainable and environmentally appropriate agricultural system.

It is unlikely that a food-labeling regime for GE foods would have disastrous consequences. It is also unlikely that our failure to implement a food labeling regime would have disastrous consequences. For the purposes of my discussion here, I set aside arguments that would focus our fears on the possibility of extreme consequences and instead focus on the rights and interests that might be affected by such a regime.

## A MELIORIST STRATEGY FOR POLICY ANALYSIS

In judging whether it would be appropriate to institute a labeling regime for GE foods, I suggest that we adopt a *meliorist* strategy: Instead of noting the ways in which the status quo fails to satisfy all relevant ideals, we need to evaluate alternative policy proposals and to judge them according to the extent to which they might constitute an improvement on the status quo. This strategy has the benefit that it directs our attention to concrete steps that we might reasonably take, instead of directing us toward an unachievable utopian ideal. For the meliorist, a labeling regime is justified only if it would respect and secure relevant rights and advance relevant interests at least as well as the status quo or the next best alternative, and as long as the costs of transition and implementation are not so great that they would counterbalance the benefits that might be achieved. In addition, a successful policy proposal should embody the least restrictive and least costly means to achieve the goals that are identified as the policy's aim.

In this chapter, I begin by considering the rights and interests that are relevant to evaluating proposals to implement a labeling regime for products with GE ingredients. I consider different institutions that might impose such a regime, and adduce arguments concerning the authority of the FDA and of state and federal legislatures to implement a requirement that GE-containing foods must be labeled.

## BY WHAT AUTHORITY?

All too often, discussions of labeling are put in broad, abstract terms. But we cannot ask whether GE foods should be labeled in the abstract. In posing the question, we need to consider specific agencies and their respective rights and obligations: What specific agency or institution (if any) should create and regulate a labeling regime that might be put in place? Different arguments and considerations apply to different agencies. Those who urge that we ought to have a labeling regime must also be specific about what this means. Should we have good reason, as citizens, to work toward the implementation of such a regime? Should legislators (which legislatures?) have a reason, or perhaps even an obligation, to frame and promote legislation that would accomplish this aim? We may be grateful to Robert Streiffer and Allan Rubel (2004) for putting the issue quite clearly and for urging that we cannot discuss the obligation to impose food labels in the abstract. Following Streiffer and Rubel, I consider separately the possibility that the FDA, state legislatures, or the federal legislature might institute a labeling regime for GE foods in the United States. For each institution, we need to consider the possibility that that institution might have an *obligation* to implement labeling, or whether it is *permissible* for that agency to do so. To say, for example, that it is permissible for the FDA to implement a labeling regime is to say that the FDA may do so without violating any obligations.

Finally, we must consider the nature of the regime that might be put in place. While there are innumerable different ways in which labeling might take place, in this chapter I consider only two alternatives: A *positive* labeling regime would require that products containing material from GE crops must be labeled as such; a *negative* labeling regime permits labeling of products that do *not* contain material from GE crops. Negative labeling might be *tolerated*, or such labels might even be facilitated and administered by regulatory institutions.[3]

This division yields 12 separate questions that need to be answered if we are interested in the issue of food labeling. Each of the 12 entries in table 10.1 corresponds to one such question. For example, entry 12 represents the question whether it is permissible for the federal government to implement a negative labeling regime. It would be a mistake to think that this table includes all of the possible questions and alternatives. Other institutions might take it upon themselves to implement a labeling regime, and we might consider in addition to the 12 questions in table 10.1 whether independent organizations may permissibly implement and enforce their own labeling regime. If we find that it is permissible for some agency to implement a labeling regime, but also permissible not to do so, we might still meaningfully ask whether there are good reasons for it to exercise that option.

**Table 10.1.** Questions about Food Labeling

| Obligation/Permission | Obligation | Obligation | Permission | Permission |
|---|---|---|---|---|
| *Positive/Negative Labeling Regime* | *Positive* | *Negative* | *Positive* | *Negative* |
| FDA | 1 | 2 | 3 | 4 |
| State Government | 5 | 6 | 7 | 8 |
| Federal Government | 9 | 10 | 11 | 12 |

Each of the 12 entries in table 10.1 represents an independent question, and a full answer to the question whether we should implement such a regime would require that we consider each of these questions. Fortunately, we do not need to consider them separately, since they are conceptually related. For example, if we were to conclude that it is impermissible for the FDA to implement a positive labeling regime, this would imply that the FDA has no obligation to implement such a regime. In this chapter I do not consider each question (each entry) separately, but instead consider only the most likely candidates. As a preview, I include in table 10.2 the conclusions I tentatively draw concerning the propriety of implementing a labeling regime.

As indicated by table 10.2, I am unpersuaded that any of the institutions in question has an obligation to implement a labeling regime, positive or negative. I note some reservations, however, about the possibility that the FDA has no obligation to implement a negative regime. While I argue that it would be impermissible for the FDA to implement a positive regime, I urge that there are good reasons in favor of an FDA-monitored negative regime. The best outcome, as I urge below, would be for the FDA to exercise this permission by creating and enforcing such a regime. As I argue, not only is it permissible for the FDA to regulate negative labeling, but also there are good moral and institutional reasons for the FDA to exercise this permission. But one might still reasonably doubt that failure to exercise this permission would not create wrongs or violate rights.

State governments are a difficult case, and I devote little attention to them in this chapter. In the United States, state governments are semiautonomous

**Table 10.2.** Tentative Conclusions about Food Labeling

| Obligation/Permission | Obligation | Obligation | Permission | Permission |
|---|---|---|---|---|
| *Positive/Negative Labeling Regime* | *Positive* | *Negative* | *Positive* | *Negative* |
| FDA | No | No? | No | Yes |
| State Government | No | No | Yes? | Yes |
| Federal Government | No | No | Yes? | Yes |

democratic regimes. They have broad latitude to legislate policies that have sufficient democratic support, and there is reason to believe that food labeling would be a policy that would attract broad support in at least some states and perhaps in most of them. There is reason for concern, however, that the federal courts might step in to prevent state governments from implementing a positive regime, because such a regime raises constitutional questions concerning free expression and interstate commerce. While I do not find these constitutional arguments to be conclusive, I note that they constitute a significant reservation concerning the right of state governments to implement such a regime. It is fully permissible, I urge, for state governments to implement a negative regime, and there are in addition good reasons for them to exercise this permission.

The U.S. Congress is quite at liberty to implement a positive or negative labeling regime, and while the constitutional reservations mentioned above could be a barrier to the actions of Congress as well as to the actions of state governments, Congress has at its disposal the democratic means to overcome any such constitutional obstacle. Whether it would or should exercise these means is another question. In particular, the fact that a majority of citizens report that they would like GE food to be labeled is not a sufficient reason for Congress to implement a labeling regime.

## WHOSE INTERESTS ARE SERVED AND SET BACK?

What interests would be served by a labeling regime, and who would be likely to oppose the implementation of such a regime on grounds that it would set back their interests?

First, there are the interests of consumers who by most accounts are overwhelmingly in favor of positive labeling proposals when they are asked about them.[4] Since people have an interest in the implementation of policies they support, a labeling regime would directly serve this interest. It is sometimes argued that labeling would serve consumer desires but not their interests, since many concerns about GE foods may be based on false beliefs or misinformation about their safety. Two considerations are relevant in evaluating this charge. First, the argument supposes that consumer dispreference for GE foods has its basis in the false belief that these foods are dangerous. While it may be true that many consumers have this fear, it is clearly not the only cognitive basis for this widespread preference pattern. If the consumer preference for non-GE alternatives were based on simple lack of information, it should be expected to disappear when consumers are provided with appropriate scientifically based information about the safety of GE foods relative to their non-GE competitor products. But, on the contrary, evidence suggests that consumer preferences do not change when they are provided with this information, and that consumers who know more—at least, those who know *a little bit* more, though who are still less than experts in biotechnology—are more, not less, skeptical about GE foods (Comstock 2000; Streiffer and Rubel 2003). The second response to the view that labeling serves consumer desires but not their interests simply insists that people must, in most standard circumstances, be regarded as the best guardians of their own interests. The attempt to supplant people's expressed interests with the interests others believe them to have is, at best, unacceptable paternalism. At its worst, it is oppressive exploitation.

Another class whose interests would be advanced consists of producers who grow or sell non-GE competitors to foods that would be labeled. On the assumption that consumers would regard the label as the mark of an economically inferior product, more consumers would move consumption to the organic and non-GE food market, increasing price and demand for non-GE products. Not all non-GE producers are small farmers: Organic agriculture is often just as heavily industrialized as traditional agriculture. But small producers often manage to stay economically afloat by catering to niche markets, and it is much easier for small producers to change their mode of operation to meet local market demand. There is thus some reason, though less than conclusive, to predict that an increase in demand for non-GE products might provide disproportionate benefits for smaller over larger producers.

Whose costs? Those likely to be disadvantaged by a positive labeling regime would include producers who grow GE crops, since demand for their product would be lower. On the assumption that a labeling regime would reduce demand for GE products, it would also be disadvantageous to people and corporations engaged in research to produce improved GE crops. One might reasonably predict, however, that only a rather small subset of the consumer population would change their purchasing habits if a labeling regime were put in place.[5] If this is so, then this provides additional support for the claim that a labeling regime would have focused benefits for the relatively smaller class of non-GE producers. The markets for organic foods and those for foods produced on small independent farms are somewhat different, since many organic farms engage in large-scale industrial agriculture.

## LABELING AND RIGHTS

A labeling regime is permissible only if it violates no rights. By parallel reasoning, the status quo, with no requirement that foods containing GE components be labeled as such, is justifiable only if it violates no rights. But where legislative chance is under consideration, rights cannot be cited in the abstract: We need an account of exactly which rights are relevant to the labeling issue, and we need a clear understanding of their scope and limits. Four rights are most commonly cited as relevant to discussions of food labeling: (1) the "consumer's right to know" what is in their food; (2) the "right to autonomy," which is regarded by some as a background right that explains why consumers have a right to know, at least in this case, what they are purchasing and consuming; (3) the "democratic right" of self-governance, which has sometimes been cited as a reason why citizens have a right that their legislators put in place a labeling regime; and (4) the "right of free expression," usually understood to include the right not to express views with which one disagrees. This last right has sometimes been cited as a reason why producers cannot be forced to state that their product contains GE elements. In an interesting case, U.S. federal courts referred to this right in overruling Vermont's state labeling law for milk produced by cows given the growth hormone recombinant bovine somatotropin (rBST).

## Consumer's Right to Know

Advocates of labeling often assert that consumers have a right to know what they are buying. Consider the following from the Consumer Union:

> Consumers have a fundamental right to know what they eat, and federal officials should require that all foods containing genetically engineered ingredients be labeled as such, including milk with recombinant bovine growth hormone. Regulatory precedent favors labeling.... [I]f the agency says that consumers should know whether their orange juice is fresh or from concentrate, why shouldn't they know about genetically engineered food? (quoted in Goldman 2000)

United States law contains no legal provision guaranteeing consumers the right to know the components of the food they eat. For example, consumers who purchase products guarded by trade secrets do not have a right to know the recipe for the product they consume. So if such a right exists, it must either be a *moral right*, or it must be a legal right that is implicit in other rights that are legislatively protected. A consumer right to know might be analyzed in different ways. In one sense, a right to know what one is purchasing might simply imply that one has a right not to pay for something unless one's questions about it have been answered. This interpretation of the right to know provides no special protections for consumers, but neither does it restrict their liberties. On a second interpretation, a right to know might be understood to imply that it is impermissible to sell an item unless one discloses everything one knows about it. This interpretation places a significant burden on sellers and may imply burdens for purchasers, as well, since the associated obligations might prevent people from placing some items on the market at all.

A third interpretation of a consumer right to know might imply that sellers have an obligation to disclose certain things that are true about the products they sell and would forbid sales contracts that did not include disclosure of these truths. Like the other interpretations, this interpretation could impose restrictions on purchasers as well as sellers, since purchase would be forbidden where disclosures were not made. But where sellers do not know the relevant properties of their merchandise, this interpretation would prohibit sale even if the seller has disclosed everything she knows about the item, and even if the buyer is willing or eager to take any associated risks. Secondary effects of such a regime could also be disadvantageous for consumers as well as sellers, since it might result in fewer items on the market and might cause an increase in the price of those items that are on the market because of the costs associated with gaining and disclosing information. The issue is relevant to the case at hand, since those who sell processed foods are often ignorant of the contents and ingredients and may have no idea whether the product they sell contains ingredients from GE sources.

A less restrictive and more reasonable solution would simply require that sellers frankly admit when they do not know whether the products they put on the market do or do not contain ingredients from GE sources. The proposal that such items bear labels indicating that "this product may contain GE ingredients" would serve this purpose.

Whether consumers may plausibly be understood to have a right to know the properties and ingredients of the items they purchase will depend on how this right is articulated and on precisely which claims, liberties, and powers this right is understood to contain. It is worth emphasizing that the "right to know" could be elaborated in ways that would restrict the liberties and claims of consumers as well as sellers and that it might be disadvantageous to both. If the "right to know" is simply an assertion that sellers have an obligation to disclose the presence of ingredients from GE sources, then we cannot simply assert a priori that consumers possess this right. If they do, then this right must be defended as the *conclusion* of an argument about the different morally relevant considerations that come in to the labeling debate; it cannot be inserted as a premise in such arguments (Buchanan 1987: 566–567).

## The Right of Free Expression

Courts in the United States have ruled that food labeling can violate the right of free expression. Constitutional guarantees of free expression include a right not to speak as well as a right to speak. If producers are required to put labels on their products identifying them as containing GE ingredients, one might regard this as a requirement that producers *affirm* the message communicated by the label. In *International Dairy v. Amestoy* (1996), the U.S. Second Circuit Court found that Vermont's statute requiring labels for milk that came from cows treated with rBST violated the free expression rights of producers, who did not want labels on their product. The court ruled that the Vermont statute (Ct. Stat. Ann. Tit. 6, § 2754(c)) violated producer's right not to speak. We should not assume that the court was right about this: It is by no means obvious that labeling a product involves any implied accession or expression by the seller or the producer. But in a full discussion of this issue, it would certainly be necessary to consider whether a labeling regime would raise a conflict with the right of free expression.

## Democratic Rights of Citizens

Democratic rights include our right to vote and our right to institutions that are appropriately responsive to democratic expressions on the part of citizens. Democratic rights, however, are typically more restrictive since only *citizens* of a state are typically regarded as having a right to have an opportunity to express their preferences by exercising the right to vote. Democratic rights are best understood as rights to institutions of a certain sort: institutions that are responsive to democratic processes and that reflect collective decisions that are made, in some sense, in accordance with the will of the majority (Rubel and Streiffer 2005: 83; Rawls

1999, 196–197; Rawls 2001). The democratic rights of citizens do not include a general right that citizens' desires and ideals must be satisfied, or even a right that preferences or ideals shared by a majority of citizens must be satisfied by public institutions. But it does include citizens' right to vote, and a right of access to public institutions including legislative institutions, which must be appropriately responsive to public preferences and ideals.

## The Right to Autonomy

In one sense, it is misleading to identify the right to autonomy as simply a right, alongside other rights we might enumerate. Just as Aristotle's "complete virtue of justice" includes all of the other virtues discussed in the *Nicomachean Ethics*, the general right to autonomy is sometimes understood to be a general right that encompasses all the other rights we possess.[6] To possess a right, in one sense, is to have a claim to make one's own decisions within the context where one's rights rule supreme.[7] In this sense, all fundamental rights may be understood as claims that protect individual autonomy.

In its most general sense, autonomy refers to the ability to direct one's own life with one's own decisions and choices. It is regarded by many theorists to be among the most basic and fundamental human values and the most basic and fundamental of all human rights. While courts have found no specific right to autonomy in the Constitution, many scholars understand a basic right to autonomy to be implied in the Constitution. There are different ways in which one might understand the concept of autonomy, and this is not the place to engage in a full-scale conceptual analysis of this important term (see Dworkin 1988; Feinberg 1986: Ch. 18). But it is relevant to distinguish two important senses of autonomy, since the concept has been cited as a basis for the other rights mentioned above. For our purposes here, I distinguish one sense of autonomy as a *capacity* and contrast that conception with an alternative conception of autonomy as a public right, guaranteed by principles of justice.

### Autonomy as a Capacity

Different individuals can direct their lives to different degrees. While public institutions can significantly influence people's ability to do so, some limitations on people's capacity to direct their lives are simply beyond the purview of public right. For example, public institutions can protect people from discrimination and can guarantee that appropriate accommodation is made for people with disabilities or other special needs. But it is not possible to ensure that everyone's life will be maximally autonomous, and there are some barriers to autonomy that the state has no authority to prevent. Further, people have a right to restrict their own autonomous capacity by making contracts, by voluntarily taking on risks, and by taking on legally enforceable noncontractual obligations. In this sense of the term, "autonomy" is a matter of degree, and it is not always the business of the state to interfere when people's choices result in limitations on their own subsequent autonomous capacity.

### Autonomy as a Public Right

Liberal states do, however, guarantee autonomy of a more restricted kind, as specified by a public conception of justice and secured by society's basic public institutions. On this conception, autonomy is a *threshold* concept, possessed when people's rights are effectively protected and when public institutions effectively regulate the contracts and legally enforceable obligations people voluntarily undertake. When we speak of "violations of autonomy," we must be referring to the threshold conception and not to the conception of autonomy as a capacity.

### Least Restrictive and Costly Means

One might argue that citizens have a general right that public institutions should use the least restrictive and least costly means to achieve legitimate state goals and to protect other rights. Public policies usually involve restrictions on liberties, and where public institutions adopt unnecessarily restrictive means, they restrict liberties unnecessarily. But our institutions have no right to restrict our liberties unnecessarily, and citizens have a right against such restrictions. Similarly, when public institutions pursue legitimate interests in a way that involves excessive costs, this can also be a violation of rights, especially if those costs are private costs and are focused on a particular group.

This very general right is important for the discussion of food labeling, since positive and negative labeling regimes would have very different costs and different degrees of restriction associated with them. A negative labeling regime, which simply helps producers to identify their product as "organic" or as "not containing GE ingredients," would provide opportunities but no restrictions, since no one would be obliged to place such a label on their product. On the other hand, a positive labeling regime would involve coercion, since it would embody a requirement that conventional producers label their product, even though they would (presumably) prefer not to do so. If one could argue that a negative regime would be sufficient to protect consumer rights and to promote legitimate public interests, then one could argue that a positive labeling regime is unnecessarily restrictive and that it violates citizens' "least restrictive means" right. If such a case could be made, it might also follow that a positive regime involved excessive costs, since more coercive policies are typically more costly than their less coercive alternatives.

## THREE POTENTIAL PROPOSALS FOR MINIMAL POSITIVE LABELING

In the sections that follow I consider three different proposals, each of which corresponds to one of the boxes in table 10.2. I hope that each of these proposals is sufficiently precise and detailed to be evaluated in meliorist terms, as a proposed improvement on the status quo. The arguments relevant to their evaluation, however, have implications for other proposals one might describe, corresponding to the unexamined questions diagramed in table 10.1. While I do not claim to have provided answers for all of the questions relevant to food labeling, I do explain some of the implications of my discussion for the unexamined questions.

I first consider whether any of the three institutions mentioned above (the FDA, state governments, or the federal government) has an obligation to institute a minimal positive labeling regime:

1. Should the FDA implement a positive labeling regime, requiring that foods that contain or might contain GE ingredients bear a label that specifies "This product may contain genetically engineered ingredients"?
2. Should state legislatures implement a positive labeling regime, requiring that foods that contain or might contain GE ingredients bear a label that specifies "This product may contain genetically engineered ingredients"?
3. Should the federal government implement a positive labeling regime, requiring that foods that contain or might contain GE ingredients bear a label that specifies "This product may contain genetically engineered ingredients"?

Positive labeling regimes may significantly vary in their strength. For example, the labeling regime envisioned by the authors of Oregon's Measure 27 would have required very specific information specifying exactly what genes and procedures had been used. The regime that would have resulted from passage of Measure 27 would have been significantly stronger than the labeling regime that exists in the European Union, since it would have required labels for products from animals that had *consumed* GE feed as well as for foods that have GE ingredients. A label that simply specifies that a product "may contain" GE ingredients would be much less restrictive and would solve some problems that would arise under a stricter positive labeling regime; for example, producers often do not know all specific features of the ingredients they use in making their products. It is not certain that such producers would have been able to market their products in Oregon at all, had Measure 27 passed.

The "may contain" regime envisioned in the three questions above may be the most minimal form that a positive labeling regime might take. It is useful to consider this minimal regime, since the conclusions we draw from this examination have further reaching implications. For example, if (as I urge) there are good reasons to think that the FDA should not impose even such a minimal regime, this would almost certainly imply that the FDA should not impose a stricter regime such as the one specified by Oregon's Measure 27.

After considering these three questions, all of which involve a *positive* labeling regime, I then briefly consider the prospects for a negative labeling regime and the questions and reasons one might use to evaluate and justify such a regime.

## SHOULD THE FDA IMPLEMENT A POSITIVE LABELING REGIME?

Among those who favor mandatory labeling in the United States, most regard the FDA as the institution best suited to enforce or regulate such a regime. Streiffer and Rubel (2004), for example, argue among other things that the FDA should understand its central mission to be the promotion of citizen autonomy, that autonomy would be promoted by a positive labeling regime, and that consequently, the FDA should oversee the administration of such a regime. Existing FDA activities, they urge, are best explained and rationalized by this assumption, since all of the appropriate regulations the FDA now enforces serve to protect and increase citizen autonomy (Streiffer and Rubel 2004).

Existing FDA regulations generally fall in two categories: Some protect consumers from potentially harmful products, while others protect consumers from false or misleading packaging and advertisement. Since protection from harm and deception both serve to promote autonomy, one might see the mission of the FDA as intimately tied to autonomy promotion. As the FDA has traditionally interpreted its mission, it is an agency charged to protect consumers from risk and fraud, by ensuring that products are safe and that labels do not contain false or misleading information. But both of these aims promote citizen autonomy: The autonomy of consumers is undermined if they are harmed by dangerous products, or if they are bamboozled by false or misleading labels. Since traditional FDA regulations all serve to promote autonomy, one might urge that the FDA should take "the promotion of citizen autonomy" as its primary mission and aim. Labels for GE foods would promote citizen autonomy, since more informed consumers will be able to act more autonomously. Therefore, one might conclude, the FDA should institute a positive labeling regime for GE foods.[8]

The preceding argument is interesting and partly persuasive. But as I argue in the following section, it should not lead us to conclude that the FDA should regard the promotion of citizen autonomy as its primary aim, or that the FDA should impose a positive labeling regime for GE foods. In fact, I argue that it is not permissible for the FDA to impose such a labeling regime. Let us suppose what may be true: that Congress, in enacting the Food, Drug, and Cosmetic Act of 1938, which brought the FDA into existence, had as its objective that the resultant regulatory institution would effectively promote consumer autonomy. Congress, we may suppose, created an agency charged to regulate products to ensure their safety and to regulate labels to ensure that the information they contain is truthful and nonmisleading. Even if this were verifiably true, it would in no way follow that the FDA should take autonomy promotion as its aim and mission. In fact, as I argue, the FDA might promote citizen autonomy *worse* if it were to take "autonomy promotion" as its mission. But first, I argue that Congress has no right to create a regulatory institution that has as its mission the promotion of citizen autonomy.

## Argument 1: Limitations on the Authority of Regulatory Institutions

One might argue that the mission of the FDA is limited by the authority that Congress could rightly exercise in creating the FDA. The power of Congress to create regulatory institutions is a limited power, and the relevant limitations are passed on to any regulatory institutions Congress may create. In particular, Congress has no right to create regulatory institutions that take over functions that should properly be served by the elected representatives in Congress themselves. To do so would be for the members of Congress to abrogate their own responsibilities by shifting them to someone else. There are good reasons why legislative positions should be subject to democratic control, while positions within regulatory institutions should be somewhat buffered from democratic forces. The reason that it would be inappropriate for a regulatory institution such as the FDA to take as its mission and charge to *promote consumer autonomy* is that this mission not only is too broad to account for the limited aims articulated in the Food, Drug, and Cosmetic Act (FDA 1938) but also is a mission that extends too deeply into realms that should be under the authority of a democratic body such as the legislature. The legislature, one might argue, has no right to create a regulatory body charged to promote consumer autonomy, because it is the job of the legislature to promote autonomy.

### Evaluation of Argument 1

How much weight should this argument receive? Clearly, the authority to create regulatory institutions is a limited power, and legislators have a responsibility to ensure that many important matters of public concern are accomplished by elected representatives, not regulatory officials. While this argument seems worth serious consideration, full development of it would require a more articulate account of the limitations of legislative authority. This is not the place to develop such a theory, and in its absence, one might reasonably find the argument less than fully persuasive.

## Argument 2: Perceived and Actual Consumer Risks

Clearly, one part of the FDA's mission is to ensure that products are not dangerous or excessively risky for consumers. But we might distinguish between the risks people perceive or believe themselves to face, and the actual risks they face. The real or actual risk associated with an option is a function of the objective probability that this activity will result in harm, and the magnitude of the harm in question. Real risks are larger when the probability of harm is greater or when the magnitude of the harm is greater. The perceived risk of an option is the *subjective perception* that the option may result in probable harm. Since we may be wrong about the probability of harm, and since we may have false beliefs about the magnitude of possible harms we believe ourselves to face, perceived risk may be significantly different from actual risk.

When people believe themselves to face significant risks, they often desire regulation to mitigate risks. But if people are wrong about the risks they actually face, they may desire regulation even where regulations will not significantly reduce the real risks they face. As consumers, we may often perceive the risks we face very differently from the way these risks are weighed by analysts employed by regulatory institutions. In such cases, people may intensely desire regulation for risks that are perceived as great but that are in fact quite small. For example, Paul Slovic's study of risk perception showed that many people had a relatively intense desire for regulation of risks associated with satellite crashes, and a somewhat less intense desire for regulation of the much greater risks associated with smoking and home swimming pools (Slovic, 1987 and 2004). This is interesting in part because the probability that one will suffer harm due to a satellite crash is *very* low—much lower than the likelihood that one will suffer harm as a result of smoking or as a result of an accident in a swimming pool. Even if a majority of people have an intense desire for regulation to reduce satellite crash risks, surely it would be a mistake for our regulatory institutions to spend extensive resources on these risks. Regulatory institutions should regulate *real* risks, not perceived risks. And where the public perception of risk is inappropriate, our institutions should educate us about them instead of simply giving us what we think we want.

This argument has direct application to the food labeling issue. Slovic's (1987) research also found that people regarded DNA technology as risky and that there was an intense desire for regulation of this technology. Concerns about GE foods have surely fed, to some extent, on the perception that they are unfamiliar and may be dangerous. But like concerns about satellite crash risks, the perceived risk of GE foods almost certainly outstrip the actual risks people face. There is no reason to believe that consumers face higher risks when they consume products with GE ingredients than when they consume organic products, and it is sometimes argued that organic products are the ones associated with higher risk.[9] Following the reasoning used in the case of satellite crash risks, one might reasonably urge that regulatory institutions should regulate *real* risks, not perceived risks. Net consumer risk would increase if our regulatory institutions shifted money from regulation aimed at reducing the actual probability of harm to regulation of more minimal risks that are associated with public fear and dread. Such a shift in regulatory focus might be popular, since it would give people regulation they want. But it would result in an increase in probable harms people would actually face. For these reasons, regulatory institutions should function very differently from democratic institutions. We should not design regulatory policies around popular fears and desires, but around actual risks and dangers. Institutions such as the FDA should therefore be insulated somewhat from the control of democratic institutions: While politicians, at least if they want to be reelected, have reason to provide people with what they think they want, regulatory institutions should be able to respond to real risks and should not be subject to manipulation from consumer fears.

One might think that citizen's autonomy is well served when public institutions give people what they want, but regulatory institutions don't work this way: Even if our reason for creating regulatory institutions such as the FDA is for the purpose of promoting citizen autonomy, it would not follow that the FDA should take autonomy promotion to be its express aim or its mission. Autonomy is undermined when people face excessive risks and when we systematically misunderstand the risks we face. We will not promote citizen autonomy if our regulatory institutions cease to regulate real risks and instead regulate *perceived* risks. Paradoxically, regulatory institutions are likely to function best to protect autonomy if they do not understand autonomy promotion to be their mission or their aim.

### Evaluation of Argument 2

It seems appropriate to distinguish between actual risks and perceived risks and to urge that regulatory institutions such as the FDA should regulate the former, not the latter. But it is worthwhile to note that these regulatory institutions must not be *entirely* insulated from the democratic process, and there is good reason to be concerned about the proposal to replace conventional judgments of risk with expert judgments by professional analysts. For one thing, professional risk analyses reflect the biases and the perspective of those who perform them. For this reason, risk analysts within an industry generally rate risks lower than external analysts who are viewing industry risks from the outside. Since analysts within regulatory agencies often come from the industries that are to be regulated, there is reason for concern that regulatory agencies may systematically underestimate the risks that people face. The problem of institutional capture, in which regulatory agencies come to represent the interests of the industries they are supposed to regulate, is a general problem, and charges of institutional capture have regularly been leveled against the FDA. There are reasons to question any risk assessment and to be suspicious when others tell us that the risks we face are very different from those we believe ourselves to face. Still, it seems right to urge that regulatory agencies should avoid replacing the assessment and regulation of risk, which is certainly part of their mission, with public relations. This is what our institutions would do if they were to begin regulating perceived risk instead of actual risk.

As an argument for the claim that the FDA should not put in place a positive labeling regime for GE foods, this argument is incomplete for several important reasons. First, people's desire for labeling may be only partly based on their perception that GE foods are risky. People may instead desire food labels because they are interested to purchase food from nontraditional producers, because they oppose industrial agriculture, or for a variety of other reasons that may have nothing to do with their perception that GE foods are risky. Second, the FDA is charged to guarantee that food labeling is truthful and nonmisleading. We must consider whether this charge would justify the FDA in imposing a positive regulatory regime like the one identified above.

## Argument 3: GE Foods and Misleading Labels

It has sometimes been argued that the absence of labels on GE foods is systematically misunderstood by consumers and that such foods are therefore misleadingly labeled. There is some support for this claim: Most U.S. consumers apparently believe that they have never eaten GE food products and are surprised to find that 70% of foods sold in American grocery stores contain GE ingredients. The fact that consumers are surprised to discover this shows that consumers are not well informed about GE foods and that they may be misled by the fact that these foods are not labeled.[10] But the fact that people are surprised does not by itself show that the labels are misleading.[11] To be culpable for misleading others, one must intentionally lead them astray, but there is no reason to believe that producers whose GE-ingredient-containing products are unlabeled are intentionally misleading consumers. Still, it is within the authority of the FDA to ensure that consumers are not systematically misled. This provides some reason in support of the claim that the FDA has a right, and perhaps even an obligation, to institute labels for GE-containing products.

### Evaluation and Response to Argument 3

Unlike arguments 1 and 2, this argument would lead us to conclude that the FDA may indeed have an obligation to institute a labeling regime for GE foods. But meliorism requires that we consider whether the labels that would be used under such a regime would be less misleading than labels that are presently in use. It is far from clear that the labels would pass this test. The FDA has staunchly resisted political pressure to require labels for foods where those labels are not associated with risks. For this reason, people may reasonably take an FDA requirement that some ingredient be identified with a special label to be an indication that there is increased risk. This expectation would be thwarted if the FDA were to part from its existing policy by requiring a special label for GE foods. Thus, the *existence* of a label might be just as misleading as its absence, since consumers might reasonably take the requirement that GE foods be labeled as an indication that GE foods are risky.

The argument that labels for GE foods would themselves be misleading is often used by pro-biotechnology advocates whose reasons for using this argument may be suspect. Still, there is reason to restrict the jurisdiction of the FDA to a strict regulatory function so that consumers *can* use the existence of an FDA requirement as an indication of risk. Since the status quo position of the FDA is that labels are not required in the absence of risk, we need good reason to believe that the alternative would be an improvement. While readers must make this judgment for themselves, I am unpersuaded that labels for GE foods are required in order to ensure that packaging is not misleading.

### Argument 4: The FDA Should Institute Labels Because They Are Required by Democratic Autonomy

When asked, a majority of U.S. citizens report that they want GE foods to be labeled. It might be urged that the FDA should institute labels because this would be a way to democratically respond to what most people want.[12]

### Response to Argument 4

I argue above that regulatory agencies should be insulated from democratic forces, because they are better able to manage public risks if they are in a position to respond to the risks people *actually* face as opposed to the risks that people perceive themselves to face. This argument would imply that the FDA should not impose regulation merely because it is desired. But this argument should lead us to consider whether democratic institutions such as state and federal legislatures might have reason to take over where the authority of the FDA runs out.

## SHOULD STATE LEGISLATURES IMPLEMENT A POSITIVE LABELING REGIME?

Legislative bodies have quite a different function and mission from regulatory institutions, and legislators have good reasons to be responsive to people's preferences, even in cases when their preferences might be thought to diverge from (what others perceive to be) their interests. For this reason, it would be *permissible*, within the context of democratic institutions, for legislators to impose a labeling requirement for GE foods as a means to respond to citizens' desire for such a regime. The sense in which it is "permissible" is a weak one: There would be no institutional impropriety in the behavior of democratically elected legislative bodies that act to put in place a positive labeling requirement as a means to respond to electoral pressures.

State legislatures may not be the appropriate bodies to undertake this project, however. If different states adopt different requirements, this could be unnecessarily costly for producers, who would surely pass these costs on to consumers. When Oregon's Measure 27 came before the electorate, the FDA urged that this legislation would impede "the free flow of commerce between the states" (Crawford 2002). If this claim is true, it would follow that the federal government has the legal power to prevent states from implementing a labeling regime. This would run afoul of the principle that regulation should be accomplished using the least restrictive and least costly means available. In any case, the cost of state-by-state labeling provisions would be high, and it would be complicated for producers to comply with a wide variety of different statutes. For this reason, it would seem that if any legislative body should institute food labeling, it should be done at the federal level.

## SHOULD THE FEDERAL LEGISLATURE IMPLEMENT A POSITIVE LABELING REGIME?

Federal legislators are positioned to be responsive to voter concerns, and federally mandated labeling for GE foods would not create special problems for interstate commerce. Federally mandated regulations have the additional benefit that they fall equally on producers in all states and thus avoid putting some producers in an unfair position of competitive disadvantage relative to other conventional producers. Since a labeling regime would be a regulation of interstate commerce, and since Congress has an express right to regulate interstate commerce, it seems clear that the houses of Congress have a right, if they choose to exercise it, to require labels. In consequence, it would seem that federal legislation would be the best way to implement a regime for labeling GE foods. If such a regime is to be put in place, the federal legislature is the appropriate institution to accomplish this.

Still, several considerations may call into question the claim that the state or federal governments should put a positive labeling regime in place. First, if a labeling regime would be expensive and would require institutional support, then it should not be put in place unless there are very good reasons for doing so. Once institutions are in place, a kind of social inertia often serves to keep them in place whether they are needed or not. For this reason, Congress should generally resist proposals to implement costly regulatory institutions unless there are powerful reasons why they are needed. Second, positive labeling would involve a restriction on the liberty of conventional producers who may currently market their products without a label. Restrictions on liberty are often justifiable, but should not be put in place unless there are good reasons. It can be argued that the mere preference of the majority, independent of the reasons that underlie that preference, is at best a very light reason in favor of removing an existing liberty.

On the other hand, the reasons supporting a liberty *not* to label one's product might be considered similarly light: While federal regulations would not impose a competitive disadvantage among conventional producers whose products may contain GE ingredients, it very well might impose upon this entire class of producers a competitive disadvantage relative to organic and non-GE producers. But it is not obvious that this competitive disadvantage should be regarded as supporting a public interest of any kind. To be sure, producers required to label their product may have a *private* interest in lobbying against such labeling, but it is not clear that they would be lobbying for any interests other than their own. Labels that provide information to consumers that want that information would not violate the rights of those who would prefer to sell their product unlabeled, and would serve the interests of consumers who may want the information labels would provide. I am led to conclude that it is quite permissible, other things being equal, for federal legislators to impose a positive labeling regime, but that they should not do so unless the cost is reasonable and the benefits are important.

We should resist, however, the argument that our legislators would be violating the democratic rights of their constituents if they *didn't* put a labeling regime in place. The mere fact that a majority would like their legislators to do something is a reason but, at best, a very *light* reason for legislators to do it. If popular desire for labeling is a mere popular preference, independent of substantial reasons backing up that preference, then legislators may have reason to respond to this preference if they (the legislators) wish to be reelected. But in a democratic republic, legislators are not supposed to replace their own judgment with the judgment of their constituents. Because of this, legislators should carefully consider the *reasons* people offer for wanting GE food to be labeled and should not simply respond to voter preference. If citizens' interest in labels is strong, and as long as there are appropriate avenues for citizen action to lobby and persuade legislators to carry out the policies their constituents want, and as long as voters have the ability to vote out politicians who do not represent their values, then procedural democratic rights are secure. In the absence of compelling reasons that militate in favor of a positive labeling regime, the mere fact of public preference should not be a decisive consideration.

## NEGATIVE LABELS AS A LOWER COST AND LESS RESTRICTIVE ALTERNATIVE

If the arguments of the preceding sections are successful, it would follow that the FDA is not at liberty to impose a positive labeling regime for GE foods. While state governments have the power to impose such a regime, their right to do so is in question, and it seems quite possible that any state that attempted an experiment like Oregon's Measure 27 would be challenged in court or by Congress.

While Congress clearly has the power to implement such a regime, some of the same considerations I have raised should cause legislators to think carefully before doing so. These conclusions, so far, are all negative. Above, I urged that we should adopt a meliorist approach as we consider whether a labeling policy is justified. It seems far from clear that a positive labeling requirement, even a minimal "may contain" requirement such as the ones examined above, would be an improvement on the status quo. In fact, I think there is an improved policy to be defended and that this policy would be cheaper and less restrictive than the minimal positive labeling regime discussed above.

Negative labels would be labels that identify products that do *not* contain GE ingredients. It would be an advantage for organic and alternative producers to be able to place such labels on their products, and when such labels are used, it would be within the authority of the FDA to ensure that the information they contain is true. Since the use of such labels would be voluntary on the part of producers, enforcement costs would be minimal. If there is a significant market for "non-GE" alternatives, then the interests of consumers who prefer to purchase these alternatives would be served if the FDA were to encourage such labels and to guarantee that the information they contain is accurate. Since negative labels would not create trade barriers, states could facilitate meaningful negative labeling regimes without running afoul of free interstate commerce.

It has sometimes been urged that negative labels would be insufficient, since they would not educate consumers about the extent to which their diets contain GE ingredients (Rubel and Streiffer 2005: 78–79). Since consumers systematically make unjustified assumptions about unlabeled food—the assumption that such food does not contain GE ingredients—one might urge that positive labels are necessary to protect consumer autonomy. But in order to support consumer autonomy, it is not necessary to ensure that people make choices that comport with their expressed values. Rather, respect for consumer autonomy requires that people have access to the information they need to make informed choices. Negative labels provide people with necessary information. Those who care about this information will look for it. If some consumers would make different choices but do not care enough to look, then there is reason to question whether their interest in avoiding GE products is a strong interest that needs federal protection.[13]

## CONCLUSION

I argue here that it would not be permissible for the FDA to impose a positive labeling regime for foods that contain or may contain GE ingredients, and that it would be similarly inappropriate for state legislatures to impose such a regime. While it is permissible for the federal legislature to require positive labeling, there are good reasons to recommend negative labels instead, as a less costly and less restrictive alternative. I began this chapter with a brief discussion of Oregon's Measure 27, which would have implemented an exceptionally strong labeling regime for the state of Oregon. How should this discussion inform our view of that measure?

While I urged that state-by-state regulation would be inefficient, this argument should not be persuasive to people who believe that the federal government is too slow to act, or to those who believe that the FDA's actions in this matter reflect institutional capture by the biotech industry. Still, there is reason to worry that Oregon's measure 27 would have been an excessive and excessively costly measure that would have done little to protect consumers. Those interested to lobby for public action in this regard would do well, from the strategic point of view, to begin with a much more modest effort. A negative labeling regime that simply provided organic and non-GE producers with a more meaningful way to distinguish their product might be a good first step that would represent a concrete improvement over the status quo.

## Notes

1. All quotations in this paragraph are from Oregon Measure 27, November 5, 2002. The text of this measure, along with arguments for and against it from interested parties, are available on the State of Oregon Secretary of State website at http://www.sos.state.or.us/elections/nov52002/guide/measures/m27 (accessed 10 April 2007).
2. The Genetically Engineered Food Right to Know Act was reintroduced in 2006 as H.R. 5269. While it has not passed, there is good reason to expect that its supporters may reintroduce it once again in 2007.
3. Philip Peters and Thomas Lambert (chapter 9) provide a helpful continuum of different dispositions that regulatory agencies could adopt with respect to the problem of food labels. For additional discussion of this issue, see Carter and Gruere (2003), Durant and Legge, (2005), FDA (2001), Gaskell et al. (2004), Hansen (2003), and Rippe (2000).
4. Streiffer and Rubel (2005) cite a variety of different studies that purport to show that people overwhelmingly desire labels for GE containing products.
5. Kalaitzandonakes, Marks, and Vickner (chapter 7) provide support for the view that most consumers would not significantly change their behavior in the presence of labels.
6. E.g., Rubel and Streiffer (2005: 82–83) represent democratic rights as a species of autonomy right.
7. The notion of rights as zones of autonomous choice is sometimes contrasted with the conception of rights as protections for fundamental interests. These conceptions may in turn be compared to Hohfeldian conceptions of rights, which may entirely bypass the dispute between choice and interest conceptions of rights.
8. I intend this to represent one aspect of the position defended by Streiffer and Rubel (2004). Since their view is sophisticated and complex, I am not confident that the objections I raise to this view are decisive against the more sophisticated view they have developed. The position described, however, is one that has significant appeal in its own right.
9. See, e.g., Pence (2001: chap. 1).
10. Streiffer and Rubel (2004) urge that the absence of labels for GE foods is misleading and that the FDA therefore has an obligation to require labels.
11. This point was made by Markie (2005).
12. A version of this argument appears in chapter 5 by Streiffer and Rubel and in Streiffer and Rubel (2004).
13. This argument should be understood to be limited: If people would care enough to look for labels if they knew more, and if they are not well educated because there have been insufficient efforts to educate people, then the fact that people do not look for negative labels will not imply that it is morally unproblematic when they purchase products they would

shun if they were better educated. In such cases, opportunities for public education should be improved. But the argument for public education concerning GE foods is not by itself an argument for food labels, since it is not at all clear that such labels are the appropriate way to educate people about GE foods.

## References

Buchanan, Allen. 1987. "Justice and Charity." *Ethics* 97: 558–575.

Carter, C.A., and Gruere, G.P. 2003. "Mandatory versus Voluntary Labeling of Genetically Modified Food, Consumer Choice, and Autonomy." *AgBioForum* 6(3).

Comstock, Gary L. 2000. *Vexing Nature: On the Ethical Case Against Agricultural Biotechnology*. Norwell Mass.: Kluwer Academic Publishers Group.

Crawford, Lester. October 4, 2002. Letter to Oregon Governor Kitshaber, www.bio.org/local/foodag/Kitzhaber.pdf (accessed Dec. 10, 2005).

Donohoe, M.T. 2003. "Politics, Pseudoscience, and Corporate Cash: The Defeat of Oregon's Measure 27." Paper presented at the American Public Health Association 131st annual meeting, San Francisco, CA.

Durant, R.F., and Legge, J.S. 2005. "Public Opinion, Risk Perceptions, and Genetically Modified Food Regulatory Policy." *European Union Politics* 6(2): 181–200.

Dworkin, Gerald. 1988. *The Theory and Practice of Autonomy*. Cambridge: Cambridge University Press.

FDA, 1938. Federal Food, Drug, and Cosmetic Act. Available at http://www.fda.gov/opacom/laws/fdcact/fdctoc.htm. Accessed 10 April 2007.

FDA (Food and Drug Administration). 2001. Guidance for Industry: Voluntary Labeling Indicating Whether Foods Have or Have Not Been Developed Using Bioengineering, www.cfsan.fda.gov/~dms/biolabgu.html.

Feinberg, Joel. 1986. *Harm to Self*. New York: Oxford University Press.

Gaskell, G., Allum, N., Wagner, W., Kronberger, N., Torgersen, H., Hampel, J., & Bardes, J. 2004. "GM Foods and the Misperception of Risk Perception." *Risk Analysis* 24: 185–194.

Goldman, K.A. 2000. "Labeling of Genetically Modified Foods: Legal and Scientific Issues." *Georgetown International Environmental Law Review* 12: 717–760. Available online at http://findarticles.com/p/articles/mi_qa3970/is_200004/ai_n8898627. Accessed 10 April 2007.

Hansen, K. 2003. "Does Autonomy Count in Favor of Labeling Genetically Modified Foods?" *Journal of Agricultural and Environmental Ethics* 17(1): 67–76.

*International Dairy Foods Association v. Amestoy*. 92 F. 3d 67 (1996).

Markie, P. November 4, 2005. "Mandatory GE Labels and Consumer Autonomy." Presentation at a meeting on Genetically Engineered Food, Columbia, Mo.

Oregon Measure 27, November 5 2002. Available on line at http://www.sos.state.or.us/elections/nov52002/guide/measures/m27. Accessed 10 April 2007.

Pence, G. 2002. *Designer Food*. Boston: Rowman & Littlefield.

Rawls, J. 1999. *A Theory of Justice*, rev. ed. Cambridge, Mass.: Harvard University Press.

Rawls, J. 2001. *Justice as Fairness*. Cambridge, Mass.: Harvard University Press.

Rippe, K.P. 2000. "Novel Foods and Consumer Rights: Concerning Food Policy in a Liberal State." *Journal of Agricultural and Environmental Ethics* 12: 71–80.

Rubel, A., and Streiffer. R. 2005. "Respecting the Autonomy of European and American Consumers: Defending Positive Labels on GM Foods." *Journal of Agricultural & Environmental Ethics* 18: 75–84.

Slovic, P. 1987. "The Perception of Risk." *Science* 236: 280–285.

Slovic. P. 2004. *The Perception of Risk*. London: Earthscan Publications.
Streiffer, R., and Rubel. A. 2003. "Choice versus Autonomy in the GM Food Labeling Debate." *AgBioForum* 6(3): 141–142.
Streiffer, R., and Rubel, A. 2005. "Democratic Principles and Mandatory Labeling of Genetically Engineered Food." *Public Affairs Quarterly*. 18(3):223-248.

# 11

# Different Conceptions of Food Labels and Acceptable Risks

## Some Contingent/Institutional Considerations in Favor of Labeling

*Carl Cranor*

Transgenically modified plants (which I refer to as "GM" [genetically modified] plants for short) result from the isolation and insertion of genes into plants by means other than conventional breeding techniques (NRC 2002: 43). Consequently, it is possible to isolate a gene that controls a valuable trait without also inserting thousands of other genes, as can be the case in conventional breeding from species that are not sexually compatible with the recipient (horizontal gene transfer). Genetically modified plants not only can confer pesticide resistance to a food plant; they can also be created to produce additional nutrients or possibly enhanced vitamin content (e.g., so-called vitamin A–enhanced plants). These new technologies are not without possible risks, and the public may have considerable concern about the wisdom of the technology or acceptability of any risks they might possess.

The Food and Drug Administration (FDA), charged in part with ensuring the safety of the nation's food supply, has concluded that foods created by GM processes (instead of Mendelian genetic breeding) will not be labeled as being from GM procedures as long as the GM foods are "substantially similar" to nontransgenically created foods. More significantly, although the agency arguably has the authority to require thorough premarket review of GM food products under the Food, Drug and Cosmetics Act of 1938's grant of authority to regulate "adulterated foods" and "food additives" (Food, Drug and Cosmetic Act of 1938), in practice the FDA has conferred a "generally recognized as safe" (GRAS) exemption from premarket review on any GM food crop deemed substantially equivalent to its traditionally bred parental strain. Moreover, the FDA has allowed product manufacturers themselves to determine whether a substance should be categorized as GRAS (Kysar 2004, p. 559). In addition, this policy seems to presume (instead of requiring evidence) that genes that were unproblematic in one plant will be safe when transferred by GM techniques to another plant.

Consequently, for the above reasons, the FDA has ruled against the mandatory labeling of GM foods. At the same time, it has also ruled that no seller may voluntarily label nontransgenic food or foods having no GM components unless the manufacturers can ensure that their products are "completely free" of GM ingredients. However, the agency suggests that such claims are likely to be "misleading" and thus illegal (Kysar 2004; FDA 2001). As a result, given the current legal environment, it appears that consumers are largely "boxed out" of knowing about GM foods—manufacturers are not required to label them unless they are "substantially different" from traditional foods, but consumers who seek to know whether their foods have been created by GM processes are not permitted to know by means of voluntary labeling that some foods are free or relatively free of having been created by GM. This legal outcome seems unacceptable. Such a concern is even more troubling given a variety of surveys indicating that the citizenry have a substantial interest in knowing whether their food has been created by GM techniques (Klintman 2002; W. K. Kellogg Foundation 2004; Streiffer and Rubel 2004), as well as attitudes toward risks that reflect much deeper concerns than mere perceptions about risks.

This chapter presents some "contingent/institutional arguments" for the labeling of GM-created foods. Given the other chapters in this volume, at a minimum it appears that there are quite compelling arguments for the voluntary labeling of GM foods. And, some of the chapters argue for their mandatory labeling. It is easy to concur with firms having the option of voluntary labeling of GM food. Arguing for mandatory labeling is somewhat more difficult, but that is a view with which I have some sympathy.

I consider several different "institutional" considerations—the scientific studies that inform the assessment of risks, notions of acceptable risks from the risk perception literature supplemented by philosophic analysis, and one institution, the U.S. Department of Agriculture (USDA), that is charged with ensuring the environmental safety of GM plants—as well as other contingent considerations that bear on GM food labeling. The functioning of the institutions bears on issues of labeling. They are contingent institutions with which we are familiar, but how well they perform or could be made to perform may affect whether GM foods and plants should be labeled in some way. I argue that issues concerning these institutions present an easy case for voluntary labeling of GM foods (and perhaps mandatory labeling, but I do not argue for that here). Moreover, as I argue toward the end of this chapter, the contingent features of these institutions are not so easily modified, especially the chances of individual risk studies to result in false negatives. Thus, although one might offer more a priori arguments from conceptions of moral autonomy or political philosophy and democracy such as other authors in this volume have presented, I believe these contingent/institutional arguments can be important considerations in favor of labeling.

## DIFFERENT KINDS OF LABELS/WARNINGS

I begin with a brief survey of different kinds of warnings and disclosures in order to have a broader understanding of some of the possible forms labels could take.

Disclosures of information provide quite reasonable models for labeling foods containing GM products.

There are warnings of danger, for example, "Bridge out!" However, actual dangers are not an issue with regard to GM foods, or the FDA would not permit them into the market. There are various kinds of warnings concerning risks (perhaps closer to issues raised by GM foods, but there must be risks before there can be warnings for them). Major league baseball stadiums may warn of the risk of being hit by foul balls. Cigarette packages warn that "Smoking can be dangerous to your health," or "Smoking kills!" (Great Britain), or "Cigarettes cause mouth disease!" (Canada, replete with vivid pictures). Again, if there were actual risks from GM foods, since the agency is charged with ensuring the safety of the food supply, it seems likely it could not legally permit foods with risks that it could not certify as "safe."

The labels on pesticides, such as the one on the common pesticide Roundup, provide safety information, guidance for proper use, and even attempted exoneration to shift the liability from the producer to the user of the product (these also do not seem apt for GM foods). Legally, the pesticide law, the Federal Insecticide, Fungicide and Rodenticide Act (1972), requires such labels; they are one significant means by which risks from pesticides are managed and reduced following the approval of a product for commercial use.

Disclosures of information perhaps provide more interesting models when we consider GM foods. Disclosures are different from warning labels for either dangers or risks. Disclosures tend to provide recipients with information in which they might be interested. Some of this information might be about risks or other potential vulnerabilities, but some of it can also be merely about items important to the recipient for any purpose.

Real estate sales disclosures in California constitute one elaborate kind of example. These require, inter alia, what is on the property, whether it is in good working order, any "significant defects/malfunctions" in structural parts of the house, surrounding components, and some 16 other features of the property, including environmental hazards, common property boundaries or conveniences, zoning violations, and so on. A number of the items on the disclosure list are merely or primarily informing the purchaser about the items on the property, and sometimes whether there are physical, financial, or legal risks that are part of the property. A most interesting feature is that a seller must disclose whether there have been any violent deaths on the property within the last three years.

Such disclosures require the seller to provide not merely information about risks, but also other information about the property. Some of them, such as the information about violent deaths, or features of the property that are merely issues about which a potential buyer might be concerned, could serve as one kind of model for informing a person in the market about something about which they strongly care, without their necessarily being exposed to "risks" properly speaking. There are other kinds of labels, some mandatory, some voluntary, that also seek only to inform purchasers about the process by which products were produced (or perhaps inform them about the product). These include mandatory labeling required by the FDA of artificial colors and flavors, voluntary labeling of kosher foods and organic foods, labeling of "dolphin-free" tuna (Kysar 2004), and mandatory labeling of irradiated foods (Degnan 2000). Moreover, none of these seems to be product related (as many insist must be the case for GM foods), yet they are either legally permissible or legally required. Our society, thus, appears to have developed inconsistent approaches to the labeling of products in commerce over time.

There is an often-expressed view that labels on GM products will tend to discourage their sale or use. There is probably some truth to this, if attitude surveys toward GM products are approximately correct (W. K. Kellogg Foundation 2004). However, such discouragement of purchases in the California housing market has not occurred because of mandatory disclosures concerning features of real estate for sale. Moreover, a real estate agent with whom I have worked and who supplied the above information about disclosures strongly supports full disclosures in real estate sales. She has never lost a sale because of them. In addition, disclosures and other kinds of labeling exist because there was sufficient support in the body politic to make them mandatory, or where the labels or disclosures are voluntary, there is a sufficient market for the product (e.g., organic foods) because there is sufficient consumer support to make it worth a seller's time, money, and effort to label them. Finally, as other chapters in this volume point out, there is quite a compelling case for legally permitting the voluntary labeling of foods and food products produced by nontransgenic methods.

The labeling of organic foods provides an attractive model for the labeling of GM foods. The USDA initially planned to follow the lead of the FDA and permit only product-oriented (not process-oriented) labeling of foods as "organic" (as the FDA required for GM foods and food products). Moreover, since it would only require the labeling of the end product, not the process that led to it, the agency planned to permit as organic foods those created by GM processes, foods treated with radiation, or foods grown with waste sludge. That is, these would all be considered as organic, as long as they were not cultivated by processes that were not judged to be organic, for example, by the use of pesticides or artificial fertilizers. However, after it received 300,000 emails, most of them opposed to the initial proposal (20 times greater than ordinary *Federal Register* responses), it retreated and provided for process-oriented labeling with varying degrees of organic ingredients (Klintman 2002).

A modest conclusion from this too brief survey is that labels on products or labels about processes do not necessarily need to be warnings about risks. They can be merely disclosures of information that are strongly desired by the citizens of a democracy about such things as the sale of houses, ingredients in foods, organically produced foods, irradiated food, or foods containing ingredients or created by processes that are important to significant numbers of individuals for religious or cultural purposes (Kysar 2004). Perhaps GM foods are of sufficient concern that voluntary labels should be permitted. In addition, labels or disclosures do not seem to be as threatening, harmful, or costly as some might worry that labels on GM foods or food products would be. Thus, on the face of it, there seems to be no strong institutional bar to use of voluntary or mandatory disclosures or labels about products (or the processes by which they are created) sold in commerce.

And, as I suggest below, there are various reasons to think that disclosing information about the processes by which foods have been produced can be quite helpful to the consuming public in a democracy.

## CONCERNS ABOUT SCIENTIFIC METHODOLOGIES

A second group of issues important to the debate about labeling GM foods is a set of concerns about the scientific methodologies used to judge the risks of products or the substantial similarity of foods. The scientific studies by which judgments about the risks from a technology are made tend to have some asymmetries between different kinds of mistakes that might result from the underlying studies. Considered as a whole, the scientific studies utilized to assess risks from products tend more to result in false-negative mistakes, that is, no effect studies mistakenly indicating that a product does not pose risks when in fact it does, than false-positive mistakes, that is, positive studies mistakenly indicating risks when there are none. This asymmetry in turn suggests that if the public were aware of these tendencies about scientific studies used to indicate the presence of risks from products, it might have much different concerns about the scientific processes than those who typically conduct them.[1]

One concern is whether human and animal studies utilized to assess risks from products, on balance, are more likely to mistakenly fail to identify risks when they are present (a false negative [FN]) or whether they are more likely to mistakenly show a risk as present when it is not (a false positive [FP]).

When human studies are used to identify risks from products, there are some features of the methodologies that, when results are positive, are likely to include FPs, but there appear to be more features of studies that, when results are negative, are likely to include FNs. In a study that shows a positive association between exposure and adverse health effects, social confounders and concomitant exposures should be closely examined to see whether that outcome is a FP, because these are typical errors that can occur with human studies (Grandjean 2004). However, when human studies show "no effect" or negative results, they may have inadequate follow-up or lost cases. They also can have difficulties in detecting increased adverse effects among common diseases (because increased adverse effects will be difficult to detect), in detecting subtle effects, or in identifying increases in rare diseases (when studies do not have sufficient power). They can be too small, have too short a duration, or be conducted too long after exposure has ceased to detect the risk of interest (Cranor 2006). All of these can cause false negatives in studies that show no effect. Such studies should be carefully examined to try to determine whether they produced a mistaken "no effect" outcome, but it is often quite difficult to determine when this has occurred in a study. Whether a positive study results in an FP or a negative study results in an FN can depend upon how carefully a study is conducted, but there are some endemic problems that are very difficult to overcome, especially the fundamental insensitivity (and high cost) of epidemiological studies in a variety of contexts (Cranor 1993: 29–30; Rothman 1986: 68; Cranor 2006: 97). As the joke has it, "Some risks are so large that even epidemiologists can detect them." Researchers can protect against chances of FN mistakes by conducting sufficiently large, carefully designed studies with long follow-up periods to allow for the latency period of disease, and with sufficiently high sensitivity for subtle or rare adverse effects. While such studies *can be done*, ordinarily their costs would be quite high, perhaps so high they would not be conducted. Often such large studies not been conducted and studies have not been carefully examined for FN mistakes; for example, Rothman reports a study by Freiman in which 71 clinical trials reported "no effect" results but that "in the great majority of such trials the data either indicated or at least were consistent with a moderate or even reasonably strong effect of the new treatment" (Rothman 1986: 117–118; Cranor 2006: 230).

For these reasons, it will likely be difficult for human epidemiological studies to detect subtle risks from human exposures to GM foods either before or after they enter commerce. If such studies produce a "no effect" outcome, it will be problematic to be sure that such a result is not mistakenly negative. Moreover, a standing lesson from epidemiology is that "there is no evidence of an effect" does not imply "there is evidence of no effect" (Cranor 2006). Consequently, even if studies do not show evidence of an effect, it does not follow that there is evidence of no effect until some very special conditions have been satisfied (which is very difficult to do) (International Agency for Research on Cancer 1998).

Similarly, animal studies, also used to detect risks to mammalian biology from exposure to substances, perhaps on balance have a greater tendency to produce FNs than to result in FPs, although here it is a closer call than for human studies because it is arguable that animal studies tend to be less insensitive than human studies. High doses to which animals can be exposed, perhaps use of very sensitive species, and some difficulties extrapolating from animals to humans are all factors that should be examined closely when such studies show a positive association between exposure and an adverse effect. These methodological features could tend produce FPs. However, a limited range of doses, low genetic variability in the experimental animals, and exposure to only one substance at a time in studies that result in "no effect" results need to be closely examined, if they can be, to try to determine whether they have produced FN results (Grandjean 2004). In general, scientists appear to have devoted considerable attention to and have developed the means of reducing FPs from animal studies, but it is less clear that similar attention has been devoted to preventing FNs (simply because in scientific research this appears to be a lesser concern of scientific methodology [Cranor 2006: 183–190]).

If we consider tendencies common to both human and animal studies, post hoc hypotheses in positive studies could lead to FPs, whereas trying to avoid false alarms, low statistical power, use of 5% $p$-values, and 20% or higher values for the FN rate of studies would lead to FNs. On balance, it appears that individual studies showing "no effect" in humans and animals have somewhat greater chances of being FNs than studies showing "positive" results have of being FPs, although this cannot be quantified. Thus, when studies show "no effect" results in humans or animals exposed to GM foods or products, there can be substantial chances of FNs. If researchers hope to rule out FNs, they will need to devote more attention to experimental design efforts and funding for larger and longer studies to decrease the chances of FNs. In particular, studies will need to be sufficiently large and conducted sufficiently long to detect any adverse effects with long latency periods and more subtle adverse effects.

Other potential risks from the production of GM foods are various environmental risks. These possible risks are much more realistic, and a few have been documented. Consider a limited range of these possible outcomes, namely, a few risks of environmental contamination from the use of GM plants to produce the foods in question. Both Norman Ellstrand, a plant geneticist, and a National Academy of Sciences panel have suggested some potential environmental risks from GM plants.

Ellstrand has argued that GM traits incorporated into plants tend to be dominant, because this is the only way in which new traits will successfully manifest. However, if wild relatives of domestic GM plants have been contaminated with the introgressed (inserted by genetic processes) gene, they will almost always express the transgenic trait. That is, a dominant trait in one plant is highly likely to be a dominant trait in the other, not a recessive trait. If the transferred trait is one that confers pest resistance or herbicide resistance, the wild relative, such as a weed or other undesirable plant, is also likely to posses it. Thus, introgressed transgenes may result in increased weediness or greater pest or herbicide resistance. Ellstrand (2003) reports that 31 of 53 deregulated GM plants approved by the USDA are plants created with pest or herbicide resistance.

In fact, this risk now appears more than theoretical. In Canada, three varieties of canola, two transgenic and one not, each resistant to a different pesticide, together seemed to have produced volunteers resistant to multiple herbicides. And, the genetic alleles moved quite rapidly (more than 550 meters from the putative pollen source in 17 months) (Ellstrand 2003). Moreover, Reichman et al. (2006) have reported the "escape of transgene into wild plant populations within the USA...glyphosate-resistant (Roundup resistant) creeping bentgrass...." There is, thus, documented potential for gene transfer to wild relatives with potential adverse consequences. In addition, although there are perhaps lesser chances of extinction, depletion of biodiversity, and some genetic pollution from genetic contamination from GM plants, these remain realistic, though less probable, possibilities (Ellstrand 2003).

A National Academy of Sciences report has suggested some other possible adverse outcomes. When GM plants have been approved for "nonregulated" status, namely, "that a particular transgenic plant is not a regulated article—that is, that it does not fall under the definition of a plant pest" (NRC 2002: 111), they have the potential to mate with similar wild or domestic plants and transfer their transgenes to new varieties as occurred with bentgrass, but legally it will be difficult to address the problem. In short, any environmental problems they might pose could be amplified because the plants have the potential to procreate, mutate, and migrate (Cranor 2004: 42).

An additional, but important, consideration that bears on assessing the possibility of environmental risks is the state of the sciences involved. To the extent that a technology and the environmental context into which it will be introduced are poorly understood, this provides reasons for going forward with caution and care until the relevant sciences are much better understood (Cranor 2004: 36). On this point, the Academy of Sciences report expresses a more disturbing and endemic issue about GM plants. The relevant environmental sciences needed to identify ecological risks from GM plants are not well understood. Moreover, the National Research Council (NRC) points out that genetics is in its infancy (NRC 2002: 184). Consequently, given the relative lack of understanding of the ecological sciences as well as the comparatively limited understanding of the new genetics, there is substantial concern about science's understanding of the ecological effects of GM plants. This particular concern is greatly heightened because with GM processes, "a much broader array of phenotypic traits can now potentially be incorporated into plants than was possible two decades ago." (NRC 2002: 14) To the extent that this report is correct, we in the United States appear to be acting with substantial ignorance about the consequences of our commercial and regulatory actions as these disciplines are rapidly developing.

A preliminary conclusion of this section appears to be that in the efforts to detect human health risks from GM plants used to produce foods, the studies used to determine risks are more likely to result in "no effect" studies being FNs (miss potential problems) than to cry "wolf" about nonexistent problems. For example, a decade-long GM food project in Australia "to develop genetically modified peas with built-in pest-resistance has been abandoned after tests showed they caused allergic lung damage in mice" (Young 2005: 18). An additional and more disturbing feature in this report suggests that even if a quite safe protein from one plant is transferred to another plant, subtle structural changes may change the protein from a benign, even beneficial, protein to one that can cause problems, as this one did (Prescott et al. 2005). If other studies show subtle protein changes when proteins are transferred from one plant to another, this may suggest that the GRAS exception for seemingly safe proteins in non-GM food plants should not necessarily obtain when these are transferred to other plants by means of GM processes. Much more research should be done to reduce the possibility of insensitive studies.

For environmental harms, the risks are more realistic. When scientists report "no adverse human or environmental effects" from GM foods, food products, or the plants that produce them, a reasonable person should recognize that there are legitimate possibilities of mistakes on the human health side and more realistic risks on the environmental side. A person might reasonably not rest too easily that scientists have correctly ensured that there are no significant risks.

Moreover, the public is more likely to be concerned about no effect studies being FNs than about positive studies being FPs. (It is difficult, of course, to judge how many studies will be positive and how many will be negative when scientists try to assess risks.) For public health or environmental purposes, the aim should be to strongly protect against FNs, but these are precisely the mistakes that no effect studies risk. Thus, the aims of public health and environmental protection appear to be in tension with the mistake tendencies of the scientific studies that inform them.

By analogy, it appears that there are likely to be similar tensions between public concerns, and tests or studies that aim to show GM foods are "substantially similar" to existing foods produced by normal Mendelian breeding techniques. Just as scientists are careful to avoid inferring that "there is affirmative evidence of no adverse effect" from "there is no evidence of an adverse effect," we should be concerned about the inference from "there is no evidence of substantial difference between GM and non-GM foods" to "there is affirmative evidence that there is no substantial difference between them" (this second point is much more difficult to show substantively) (Cranor 2006: 243–245). Moreover, the FDA seems to assume this point (or permit manufacturers to assume it), rather than having manufacturers establish it (Kysar 2004). This raises the following questions: To what extent should the public have high degrees of confidence that there are no or few risks from GM foods? And, to what extent should the public have high degrees of confidence that there are no substantial differences between GM-created and Mendelian-created foods and food products? Given the mistake tendencies of studies that must be utilized to determine there are no adverse effects from GM products or there are no substantial differences between GM and non-GM products, any public concerns about new technologies are not wholly misplaced.

Food labels identifying products that are the result of a GM process would assist in counterbalancing some of the above problems. If the citizens were informed and concerned enough about the mistake tendencies of the scientific methodologies or whether there are substantial similarities between GM foods and Mendelian-created foods, and had strong interests in going more slowly than companies in introducing such technologies, they could cast their vote accordingly by means of their purchases, if they knew about GM foods by means of labeling. Some citizens no doubt are like my real estate agent, who noted, "If I don't grow it, I don't eat it."

## ACCEPTABLE AND UNACCEPTABLE RISKS

Paul Slovic and various collaborators have shown over the years that there are a variety of dimensions to the public's perceptions of risks (Slovic 1987). This research tends to suggest that the public perceives the risks with properties that tend to fall in the second column in table 11.1 as worse than risks in the first column. However, there is an ambiguity in this language: It might mean that the public judges the risks with properties in the second column as more probable and more serious in magnitude than risks in the first class. It might also be taken to mean that features of risks in the second group tend to be less acceptable to the public than risks in the first group (at least somewhat independent of the probabilities involved). I believe for the most part the second interpretation of the language is the more accurate characterization of risk judgments.

Elsewhere, I have argued independently that there are a variety of good normative reasons for a citizen or conscientious moral agent to believe that risks are either less acceptable or not acceptable compared with the judgments of technical experts who base their views on the probabilities and magnitudes of the risks involved (Cranor 1995, 2007). That is, risks have normative features that affect the judgments of their acceptability that are more or less independent of their probability

**Table 11.1.** Normative Judgments about Risks

| More Acceptable Features | Less Acceptable Features |
|---|---|
| *Dimensions of risk perception* | |
| Controllable | Not controllable |
| Merely injurious | Fatal |
| Equitable | Not equitable |
| Low risk to future generations | High risk to future generations |
| Easily reduced | Not easily reduced |
| Voluntary | Involuntary |
| Does not affect me | Affects me |
| Not dreaded | Dreaded |
| *Epistemic considerations* | |
| Observable | Not observable |
| Known to those exposed | Unknown to those exposed |
| Effect immediate | Effect delayed |
| Old risk | New risk |
| Risk known to science | Risk not known to science |

From Slovic (1987)

of occurring and the magnitude of damage they might do, were they to materialize. In Slovic's (1987) language (discussed below), I would say that the public has good normative reasons for judging that risks with features in the second group in table 11.1 are less acceptable than risks in the first group[2] (Cranor 1995, 2007). Here, I briefly review some of the different dimensions of risk perception that correspond more or less closely to normative judgments that make risks more acceptable or less acceptable. Table 11.1 summarizes Slovic's (1987) findings, but with my titles—more acceptable features of risks compared with less acceptable features.

If one reviews this list of dimensions of risk perception, many of the features in the second column that tend to lead to risks being perceived as less acceptable apply to GM foods. Persons do not find them controllable (because they are usually unaware of their presence), any risks are not spread equitably between producers and risk recipients, any risks to the environment could easily extend to future generations (through genetic modification in related plants), and exposure to them is involuntary under current laws. Moreover, they are largely not observable, are unknown to those exposed, are likely to be delayed (not immediately manifesting adverse effects), and constitute new risks (if there are any), and science likely does not understand them well according to the NRC (2002) report on environmental effects of GM plants. Thus, based simply on the dimension of risk perception, the public is likely to perceive that there are risks from GM foods, or even if they do not, they are likely to be quite concerned about them because of the above properties. Moreover, various surveys confirm these concerns (Klintman 2002; W. K. Kellogg Foundation 2004; Streiffer and Rubel 2004).

Slovic's work could be and has been read as pointing out that that these public views are merely perceptions of risks. However, it seems that they are not just perceptions of risks that could be dismissed as "mere perceptions" the way one might dismiss a mirage or some other kind of mistake on the part of the observer. Although some risk theorists have appeared to argue that although the public has such perceptions and risk theorists appear to dismiss them as "mere [mistaken] perceptions," they should not be treated as mistakes about a more correct underlying reality. The issue is the *acceptability* of the risks, not whether the perception is a correct view of an independent reality. I believe that there are good reasons for conscientious agents to regard most of these dimensions as normatively well grounded.[3] Consider a few of them.

First, there is a point Slovic (1987) does not discuss: There are good reasons to distinguish between naturally created and humanly created risks. The first pose several problems, for example, how serious are the risks, how dangerous if they materialize, and what should be done about them in order to minimize them or reduce their chances of materializing? For humanly created risks, we have all the concerns that are present for naturally occurring risks, plus we could choose not to have the activities at all (although there might be substantial costs from not having them), or they could be conducted in quite different ways than they currently are. Collectively, we have much greater control over humanly created risks than we have over at least the more visible naturally created risks from hurricanes, earthquakes, and the like.[4] Thus, humanly created risks are much more plastic, malleable, and controllable (at least collectively) than are many of the more obvious naturally caused risks. Suppose that the probabilities of naturally occurring risks are identical to humanly created risks. We might judge the humanly created risks as much less acceptable, simply because collectively we have greater control over them, could rearrange the institutions that create them, and think perhaps that our institutions should be restructured to minimize the risks to fellow citizens. Thus, for example, on this dimension, if there are any risks from GM foods, these will tend to be less acceptable than naturally occurring risks of similar magnitude and probability. Humanly caused risks, if there are any, from GM foods are in principle collectively much more controllable—in the extreme, we need not have them at all.

A second dimension bearing on the acceptability of risks concerns welfare benefits and harms to the persons involved. By welfare benefits and harms, I mean features of the activity associated with the risks that tend to promote or injure a person's well-being, or "what it means for a human life to go well" (Griffen 1986: 7). While there are some generic benefits from GM foods—for example, possible cost savings to consumers, increases in production capacity to the farming community (which can indirectly benefit persons), ability to feed larger numbers of people, perhaps reduced pesticide use in production in the short term—there appear to be few direct welfare benefits to the vast majority of individuals (cost savings in GM foods would be the most obvious).

Also, for some risks there are what we might consider direct benefits of other kinds to persons that could make undertaking risks more acceptable—the activity is part of a personal project; it has important experiential features to it (makes the person's life worth living), for example, rock climbing or backcountry skiing; it calls upon one's expertise, or permits one to exhibit substantial expertise, for example, skill requiring activities such as rock climbing or race car driving; or it is a personally morally rewarding activity, for example, being a lifeguard or an emergency medical technician. Genetically modified foods offer none of these kinds of direct benefits that in other contexts tend to make risks more acceptable to the persons involved (except to the firms and scientists within them who develop the technology). Consequently, there appear to be almost no benefits to the vast majority of persons and none of the other benefits that tend to make risks from other activities acceptable to the agents involved or make the risks morally acceptable.

A related dimension is that some risks to which a person is exposed tend to be created by the same person. In effect, the risk creator is the risk recipient; for example, rock climbers by engaging in the activity impose many risks on themselves (and perhaps some on others). This internalizes the risks and tends to make them acceptable to the agent. However, if GM foods result in risks, these for the most part are not internalized to the risk creator—the company that creates the GM products (although a firm clearly internalizes some risks, e.g., financial and other possibilities of product failure). Instead, if there are risks, there are likely to be substantial externalities to third parties or to the environment (especially to the environment), potentially creating substantial distributive issues that would have to be addressed as a matter of social policy in order for the risks to be judged acceptable.

A person's voluntary embracing of a risky activity or exposure tends to make the exposure or activity acceptable (Cranor 1995, 2007). However, there are some conditions on voluntary acceptance that are not likely to be satisfied in the case of any risks from GM foods. At present, there appears to be little or no voluntary choice in the public's exposure to GM products and any contaminants they might possess, since the FDA does not permit notification that a person is ingesting a GM food or food product. They are, for example, epistemically undetectable. There is no or little awareness, palpability, or appreciation of the risks of the kind that often makes risk exposures more acceptable (or that make the risks detectable so that individuals are put on notice about their presence and can begin to exercise choices with respect to them). Unlike many of the risks of life from macro-objects such as cars on the streets, falling trees, visible mechanical devices, and the like, the properties of GM foods that pose risks (or at least concerns) are invisible. In this respect, they share the same properties with chemicals and radiation, other invisible intruders that can present risks (or even harm) at some level of exposure. Their very invisibility or undetectability is a feature that will enhance concerns about their safety, simply because persons cannot use their ordinary sensory apparatus and other means of detecting harms to protect themselves. Moreover, ordinary persons have to work very hard to be aware of exposure to GM foods or food products and even harder to avoid them, if they want to do so. Labeling such products would increase public awareness of such foods and give people the opportunity to avoid any products about which they had concerns. This would increase somewhat (but not fully) the possibility of voluntary exposure to the products, provided that labeling would be an effective way of providing knowledge of their presence and persons had reasonable opportunities to avoid them should they choose to do so. The less voluntary any exposures are, the less any risks tend to be acceptable.

Participation in a decision to create or permit a risk can be evidence that a person finds a risk acceptable or, when proper conditions are satisfied, even something of a waiver about the risks in question. However, this feature that might assist in making a risk acceptable appears to be missing in the case of many new technological risks, including GM products. Although there were surely announcements in the *Federal Register* about proposed rule making concerning GM foods, merely providing for comment periods does not provide robust public participation.

In many circumstances, continuing control over whether or not a risk materializes is an important consideration that contributes to its acceptability. For example, if I have continuing control over the risks created by my running a chainsaw, this suggests that I would tend to find those risks acceptable or at least not so unacceptable that I discontinue the activity. Moreover, if I have such control, I can choose to avoid them at any given time. Because I can opt to avoid them at any moment, this provides a reason to think that I find continuing exposure to the risks acceptable or not too unacceptable. In present circumstances there is little or no personal continuing control over any risk exposure to GM products, and none over the risks materializing, if there are any risks to materialize. Moreover, because there is no labeling of GM foods, the FDA has removed the main means by which a citizen might exercise some minimal control over their exposures to GM foods. By depriving the citizenry of this information, the agency seems to have made a conscious decision to prevent the possibility.

An additional dimension of the acceptability of risks is the extent to which the relevant scientific communities understand a technology. As I have described above, an NRC committee has gone on record that both the ecological and genetic sciences that are needed to understand environmental risks are poorly understood. If the public were aware of this conclusion, this would provide additional reasons for concern about the creation of GM plants for foods or others purposes. In addition, if the public were aware of the methodological tendencies of some scientific studies, these would provide further reasons for concern about the acceptability of risks or the similarities of foods. Moreover, Klintman (2002), Streiffer and Rubel (2004), and the W. K. Kellogg Foundation (2002) all report a variety of surveys suggesting that the public has considerable concern about GM processes and foods.

Finally, there are some asymmetries concerning GM products that may also have impacts on their acceptability, if the public were to become aware of them. Companies tend to conduct more research on the benefits of products than on any environmental and health risks. Thus, before products enter commerce and are subject to premarket review (this appears not to be the case for GM foods [Kysar 2004: 559]) or are subject to postmarket regulation (the more typical case), companies have incentives to explore and develop the benefits of products, but fewer incentives to fully understand any risks (this is especially the case when products are subject to postmarket legal requirements). Moreover, focused political constituencies favor protections for products rather than protections for public health and the environment. If the public were disquieted about the tendencies of these institutions, these considerations would add to their concerns (Cranor 1999). Thus, with these asymmetries in information and political power, the general public is doubly disadvantaged. If citizens had concerns about the extent to which a company understood the risks of its projects, they could opt not to purchase them until there was greater reassurance about their safety. At the same time, if they were concerned that firms had substantial asymmetric power to influence the approval process and to minimize an agency from discovering any risks from their products in agency hearings, labeling would provide citizens a way to express this concern.

To sum up, the comparative unacceptability of any risks from GM foods is suggested by generic psychometric research (Slovic 1987), good reasons for concerns about any risks (Cranor 1995), and some evidence from various attitude surveys (W. K. Kellogg Foundation 2004; Streiffer and Rubel 2004; Klintman 2002). If there are any risks or concerns about GM foods, they appear to be comparatively unacceptable when considered against many of the ordinary risks of life.

Perhaps, however, these results do not settle the issue. Without most of the features that would provide good reasons for people to find risk exposure to GM foods acceptable, there might be good reasons for finding exposure to GM foods acceptable if a government trustee did a sufficiently competent and legitimate job of reviewing products and protecting the public and environment from risks. While much more information is needed on this dimension, at least one of the agencies that has the task of reviewing GM plants to ensure there are no risks to the environment has been reviewed by an NRC committee of the National Academy of Sciences and found wanting.

## PUBLIC TRUSTEES

The acceptability of decisions by a trustee to protect the public from any product risks would appear to rest upon the trustee's technical reliability in understanding and identifying risks, and its social legitimacy in how protections are provided and implemented. An NCR panel has utilized such grounds to assess agency performance in the regulation of environmental risks from GM crops (NRC 2002; Cranor 2004).

There is divided legal authority to regulate GM plants. The USDA regulates development and field testing for environmental risks, the U.S. Environmental Protection Agency regulates GM products that might have pesticide residues in or on food as well as any toxic effects they might have, and the FDA is charged with assuring GM food safety. I consider only the USDA's review of how well the farm and nonfarm environments are protected from GM plants (including largely GM food plants).

The USDA has legal authority to assess "the potential effects of non-pesticidal GM plants on other plants and animals in both agricultural and nonagricultural environments" (NRC 2002: 19). The USDA's role is not to assess any risks from GM foods, but to assess farm and nonfarm ecological effects from GM plants that might produce the GM foods, such as GM canola or soybeans. I focus on the USDA role here simply because its regulatory processes have recently been reviewed by a National Academy of Sciences committee, and it represents an important step in the GM food industry that produces the end product.

The USDA has created three different levels of review depending upon the potential problems they pose: "notification," "permitting," and "petitions for nonregulated status." *Notification* is the legal process by which a proponent of the GM plant (that is not on a list of weeds) before field testing must "notify APHIS [the USDA's Animal and Plant Health Inspection Service] of its intent to release a regulated article" (NRC 2002: 104, 106, 108). *Permitting* is utilized for the "movement, importation and field testing of GM plants that do not qualify for notification," for example, plants that appear to pose greater environmental risks than those considered under notification and those that might be used for commercial products. Since permitting is an alternative to notification, it presumes that a plant is a "potential pest and requires anyone who wants to introduce it into the environment to obtain a permit" (NRC 2002: 110). Thus, notification is reserved for plants that do not appear to pose environmental problems, while the permitting process is utilized for GM plants for which such assumptions cannot be ensured. However, permitting is a seldom-used alternate to the notification process. If field testing under notification or permitting procedures reveals no adverse effects, a proponent can petition for "*nonregulated status*" (NRC 2002). This is a review during which the agency determines whether "a particular transgenic plant [should or should not be] a regulated article—that is, [whether it] fall[s] under the definition of a plant pest" (NRC 2002: 111). Moreover, it is the "sole route for commercialization of transgenic plants (e.g., sale of transgenic soybean seed) and the primary but not sole route to commercialization of transgenic plant products (e.g., when the plants are never sold but a product such as an industrial protein extracted from the plant is sold)" (NRC 2002: 111). The information is utilized to determine whether the regulated article displays any plant pathogenic characteristics; whether it is less likely "to become a weed than its nontransformed parent; [whether it] is unlikely to increase the weediness of cultivated, feral, or wild-related plants; [whether it damages] processed agricultural commodities; and [whether it may] ... cause unintended significant harm in other organisms" (NRC 2002: 112). This is the main focus of the discussion below.

Once a GM plant has been granted nonregulated status under this procedure, APHIS "cannot exercise any additional oversight on the plant or its descendants" (NRC 2002: 112). Descendants of such plants might include all plants of the same species that might receive the transgene through sexual reproduction, but may also include quite distantly related members of the same crop species.

The NRC report expressed concerns about both the technical reliability and social legitimacy of the USDA in reviewing environmental risks from GM plants, some of which are used for foods (NRC 2002; Cranor 2004). As case studies, the NRC committee reviewed six petitions for the nonregulated status for plants, and although none of the six petitions reviewed resulted in mistakes, the committee did not appear to have sufficient confidence that the current process would prevent mistakes.

There appear to be several procedural and even scientific inadequacies in agency practices and guidelines. For example, the committee found that the agency procedures "should be made significantly more transparent and rigorous by enhanced scientific peer review, solicitation of public input, and development of determination documents with more explicit presentation of data, methods, analyses and interpretations" (NRC 2002: 10). Moreover, since, as indicated by the committee, neither genetics nor ecology is a well-understood field of science (NRC 2002: 184), the review should be more rigorous when scientists are making decisions under substantial uncertainty (Cranor 2004: 40). And since decisions are made essentially in perpetuity, this only enhances the need for more careful decisions than the agency appears to be making. In short, there is a concern that decisions are too final for the level of uncertainty involved (Cranor 2004: 40). These various shortcomings primarily bear on the reliability of the agency's science, but in turn, they can affect the legitimacy of the regulatory procedures and outcomes in the eyes of the public if they come to realize that agency procedures seem insufficiently protective given the environmental risks at stake and the substantial uncertainties involved in the science.

In addition, the NRC committee has found some more generic problems that are independent of the particular regulatory authority to grant nonregulated status. It urges that there should be greater independent scientific input, peer review, and oversight than there has been to date (NRC 2002: 10). This is even more important when the agency is making precedent-setting decisions (NRC 2002: 10). It is also particularly strong on recommending the importance of the public legitimacy that comes from ensuring that the public has an opportunity to participate in and understand the regulatory decisions that affect their lives (a point noted above under the acceptability of risks) (NRC 2002: 10). And, in order to ensure the accuracy of procedures, there should be postcommercial monitoring to assess the precommercial accuracy of evaluation (a kind of quality control) (NRC 2002: 12–13).

In general, monitoring is a problem, since the NRC notes that the United States does not have good procedures in place to monitor its natural or farming environments. Present procedures appear woefully inadequate for assessing the current state of both the farming and natural ecosystems, detecting any newly arising problems, and assessing any changes in the ecosystem.

A more disturbing endemic issue is the NRC panel's assessment of the state of the ecological sciences and the new genetics noted above: So little is understood about both fields that there is substantial concern about the ecological effects of GM plants. For example, in the ecological area, farming itself exerts a destabilizing influence on natural ecosystems. Consequently, the "[p]otential ecological effects of transgenic crops, and other crops bearing novel traits, may be heightened.... This argues for a cautious approach to the release of any crop that bears a novel trait" (NRC 2002: 4). Such effects come on top of a natural environment that in a number of respects is already in fragile condition (Rodgers 1994: 8; Lubchenco 1998: 492). This particular concern is additionally significant because with GM processes "a much broader array of phenotypic traits can now potentially be incorporated into plants than was possible two decades ago" (NRC 2002:14). To the extent that this report is correct, we in the United States appear to be acting with substantial ignorance about the consequences of our commercial and regulatory actions.

Importantly, the NRC has a special emphasis on the importance of public input into the regulatory process (which we might think of as an aspect of our democracy, as well as an important consideration entering into the acceptability of any risks associated with new technologies). Without modification of the USDA procedures as the NRC committee recommends, they appear to have deficiencies on both grounds of technical reliability and political legitimacy, even though no obvious mistakes have been made to date.

The NRC report assessing the USDA's review of the environmental risks of GM plants does not reassure the public or conscientious agents about the reliability and legitimacy of decisions by this public trustee for GM products. If the public had greater awareness of this report, it might well have even more concerns about the acceptability of GM plants and foods derived from them. It is quite arguable that conscientious agents would have additional reasons for finding the use and purchase of GM foods less than fully acceptable.

## CONCLUDING ARGUMENTS FOR DISCLOSURES

There are reasons for explicit disclosures that foods have been created by a process of transgenic modification, contrary to decisions the FDA has made with regard to GM foods. I close with those that should be seen as supported by empirical/institutional reasons; other authors in this publication have addressed reasons based more directly on moral/legal principles. Disclosures of the GM process by which such foods are created can serve as correctives to some of the above issues to which I have called attention.

If consumers were concerned about the mistake tendencies of the relevant scientific studies involved (some tendency that no effect studies are more likely to miss risks than positive studies are to cry "Wolf!" about nonexistent risks), as it seems they legitimately should be, they could register their concerns by opting not to use products created by processes about whose risks they were unsure. That is, if they believe that although no studies have found risks from GM foods, it is difficult for such studies to ensure that "no effect" outcomes are correct negative studies, disclosure gives members of the public choices.

If, in fact, the public perceives risks from GM foods, if there are any, as less acceptable than many ordinary risks of life (and it appears that they have good reasons to rank the acceptability of any such risks lower than many of the ordinary risks of life), labels disclosing the processes by which the foods were created would give them a way to cast a democratic (but commercial) vote for or against the acceptability of the products that they purchase. The FDA by consciously prohibiting consumers from having access to such information has removed several means by which consumer-citizens could make legitimate choices about their use. Moreover, they might have a variety of concerns about the processes by which GM foods were produced. They might have concerns about long-term health and safety, effects on ecosystems and nontarget organisms that the USDA may or may not have good processes to identify, on the integrity of the ecosystem, and "a host of more general ethical and cultural considerations" (Kysar 2004: 590).

If there are genuine reliability and legitimacy concerns about the USDA as a trustee to protect the environment from adverse effects of GM products or if there were reasons to be concerned about the FDA's GRAS policy for proteins that are safe in one plant to be presumed safe in another, disclosures would provide the public a way to express their concerns about these issues and would provide the agencies with feedback to improve their regulatory procedures, so the public might become more accepting of their decisions. (If the public had no such concerns, they could express this by purchasing GM foods.) Moreover, providing the public with this choice would serve public health and environmental purposes by giving persons an opportunity to avoid products for which they might not have sufficient confidence that scientists or agencies had detected all significant risks. Disclosures would also provide them with choices if they had concerns about the acceptability of risks or the "substantial similarity" of foods.

Disclosures about the GM process would also provide information for consumers' decisions about whether or not to veto trustees' actions vis-à-vis GM foods at least as it affected them. Disclosures, in addition, provide a partial corrective to some of the asymmetric information and political tendencies concerning new products, especially those subject to postmarket regulation (consumers could opt away from GM foods if they have concerns). Finally, they would provide information on which citizens could act for political or institutional purposes, if they had moral, political, or environmental concerns about the direction or general social impact of new technologies.

The arguments presented in this chapter rest in substantial part on contingent institutional reasons, but to what extent might these considerations change in ways that would modify the conclusions? Issues concerning the mistake tendencies of the pertinent scientific studies are not likely to change anytime soon, barring some dramatic breakthroughs concerning tests for the safety of products or vastly increased investment to ensure larger scale studies are highly sensitive to subtle adverse effects. Concerns regarding the possibility of environmental risks could be modified depending upon better understanding of the ecological and genetic sciences, but modifications in the knowledge and sophistication of these sciences are unlikely to be quick. The public's legitimate views about the acceptability of GM processes are unlikely to change, simply because GM plants and foods have properties that, if they pose any risks, will present risks toward the unacceptable range of risks in our lives. The normative considerations will remain relatively stable, given the invisible, undetectable nature of risks, if any, from GM plants or foods compared with the more visible risks of life combined with distributive considerations involved. Procedures that could be altered fairly quickly are those agencies follow in reviewing the human health and environmental safety of GM plants and foods. Whether or not they will be revised depends upon the political will of the decision makers involved. However, since such procedures ultimately rest on the underlying scientific studies, the reliability of these studies will likely pose a barrier for some time into the future.

Should GM food labeling be mandatory, or should it be voluntary on the part of producers seeking to find a commercial niche, following the example of producers of organic foods? There is insufficient space here to properly address this issue. Moreover, others in this volume have presented some of the arguments for and against mandatory labeling. However, what is clear is that firms should have the legal authority to voluntarily label their products as "GM free" or "95% GM free" or "80% GM free" using a tiered labeling regime similar to one that exists for organic foods. At present, companies are not legally permitted to do so, and this seems a substantial shortcoming of law. Moreover, for the reasons given above and for reasons others in this volume cite, the case for voluntary labeling is easy. Citizens in a democracy should have access to information about new technologies, especially those whose potential risks are otherwise undetectable, those about which there appears to be substantial disquiet as revealed by polls, and those for which there are good reasons for finding their concerns legitimate (as the discussion here concerning acceptable risks and concerns shows). Whether such labeling should be mandatory will depend upon the persuasiveness of arguments to this conclusion from autonomy and democracy that others have presented. These should certainly be developed and pursued.

## Notes

1. The point I am making is about tendencies of *individual* studies. Some might think (as did some in the audience where these ideas were first presented) that in estimating the risks from products, the risk assessment process tends to overestimate actual risks. That point may sometimes be correct in individual assessments about risks, for example, from carcinogens. That is, there is some degree of risk, but the question is how great it is. In such cases, risk assessors and agencies tend to utilize models to characterize the risks that ensure that risks are not underestimated. However, if there is a product whose risks are not known, and if tests do not suggest a risk (contrary to tests that have already indicated a risk from exposures to carcinogens), the question is whether such tests if they are negative are truly negative or falsely negative. The concern above is that even if tests show "no effect" or a "negative" result, one should have some legitimate concern about whether "no effect" results are true negatives.
2. As I was completing an essay on trying to understand comparative risks judgments (Cranor 1995), I discovered Slovic (1987) and found our conclusions surprisingly similar, mine reached by means of normative philosophical analysis and his the result of psychometric studies of community attitudes toward risks (Cranor 1995).
3. For purposes of this chapter, I do not distinguish different kinds of normative reasons that agents might have for thinking risks are more or less acceptable—that is, whether their reasons are directly moral reasons (that appeal quite directly to substantive moral principles) or whether they are what we might think of as personally acceptable or unacceptable independent of substantive moral principles (with their moral status left somewhat unclear at this point). I address some of these issues in Cranor (2007).
4. However, it is important to note, as we have seen as a result of the hurricane Katrina, that human decisions made to prepare for or to try to anticipate the harms from a natural

force such as a hurricane can make a substantial difference in our judgments about the resulting harm.

## References

Cranor, Carl F. 1993. *Regulating Toxic Substances.* New York: Oxford University Press.

Cranor, Carl F. 1995. The Use of Comparative Risk Judgments in Risk Management, in *Toxicology and Risk Assessment: Principles, Methods and Applications*, ed. Fan, A.M., and Chang, L.W. NewYork: Marcel Dekker, pp. 817–833.

Cranor, Carl F. 1999. Asymmetric Information, the Precautionary Principle and Burdens of Proof in Environmental Health Protections, in *Protecting Public Health and the Environment: Implementing the Precautionary Principle*, ed. Carolyn Raffensperger and Joel Tickner. Washington, D.C.: Island Press, pp. 74–99.

Cranor, Carl F. 2004. Assessing Some of the Regulatory Approaches to Transgenic Plants: What Can We Learn from the Regulation of Other Technologies? *Environmental Biosafety Research* 3: 29–43.

Cranor, Carl F. 2006. *Toxic Torts: Science, Law and the Possibility of Justice.* Cambridge and New York: Cambridge University Press.

Cranor, Carl F. 2007. Toward a Non-consequentialist Theory of Acceptable Risks, in *Risk: Philosophical Perspectives*, ed. Tim Lewens. London: Routledge.

Degnan, Fred H. 2000. Biotechnology and the Food Label: A Legal Perspective. *Food and Drug Law Journal* 55: 301–310.

Ellstrand, Norman. 2003. *Dangerous Liaisons.* Baltimore, Md.: Johns Hopkins University Press.

FDA (Food and Drug Administration), Center for Food Safety and Applied Nutrition. January 2001. *Draft Guidance for Industry: Voluntary Labeling Indicating Whether Foods Have or Have Not Been Developed Using Bioengineering*, www.cfsan.fda.gov/~dms/biolabgu.html.

Federal Food, Drug, and Cosmetic Act (1938, amended 1958, 1960, 1962, 1968, 1996).

Federal Insecticide, Fungicide, and Rodenticide Act (1972, amended 1972, 1975, 1978, 1996).

Grandjean, Phillippe. 2004. Implications of the Precautionary Principle for Primary Prevention and Research. *Annual Review of Public Health* 25: 199–223.

Griffen, James. *Well-Being: Its Meaning, Measurement and Moral Importance.* New York: Oxford University Press, 1988.

International Agency for Research on Cancer. December 17, 1998. Preamble to the IARC Monographs: 8. Studies of Cancer in Humans" available at monographs.iarc.fr/ENG/Preamble/studieshumans.php.

Klintman, Mikael. 2002. Arguments Surrounding Organic and Genetically Modified Food Labeling: A Few Comparisons. *Journal of Environmental Policy Planning* 4: 247–259.

Kysar, Douglas. 2004. Preferences for Processes: The Process/Product Distinction and the Regulation of Consumer Choice. *Harvard Law Review* 118: 525–641.

Lubchenco, Jane. 1998. Entering the Century of the Environment: A New Social Contract for Science. *Science* 279: 491–497.

NRC (National Research Council). 2002. *Environmental Effects of Transgenic Plants.* Washington, D.C.: National Academy Press.

Prescott, Vanessa E., Cambell, Peter M., Moore, Andrew, Mattes, Joerg, Rothenberg, Marc E., Foster, Paul S., Higgins, T. J. V., and Hogan, Simon P. 2005. Transgenic Expression of Bean r-Amylase Inhibitor in Peas Results in Altered Structure and Immunogenicity. *Journal of Agricultural and Food Chemistry* 53: 9023–9030.

Reichman, Jay R., Watrud, Lidia S., Lee, E. Henry, Burdick, Connie A., Bollman, Mike A., Storm, Marjorie J., Kig, George A., and Mallory-Smith, Carol. 2006. Establishment of Transgenic Herbicide-Resistant Creeping Bentgrass (*Agrostis stolonifera* L.) in Nonagronomic Habitats. *Molecular Ecology* 15: 4243–4425.

Rodgers, W. H., Jr. 1994. *Environmental Law*, 2d ed. Minneapolis, Minn.: West Publishing.

Rothman, Kenneth. 1986. *Modern Epidemiology*. Boston: Little, Brown.

Slovic, Paul. 1987. Perception of Risk. *Science*, 236: 280–285.

Streiffer, Robert, and Rubel, Alan. 2004. Democratic Principles and Mandatory Labeling of Genetically Engineered Food. *Public Affairs Quarterly* 18: 223–248.

W. K. Kellogg Foundation. 2004. *Perceptions of the U.S. Food System: What and How Americans Think about Their Food.* Battle Creek, MI: W.K. Kellogg Foundation or http://www.wkkf.org/DesktopModules/WKF_DmaItem/ViewDoc.aspx?fld=PDFFile&CID=4&ListID=28&ItemID=44172&LanguageID=0.

Young, Emma. November 21, 2005. GM Pea Causes Allergic Damage in Mice. *New Scientist*, http://www.newscientist.com/channel/health/dn8347-gm-pea-causes-allergic-damage-in-mice.html.

# 12

# Using Food Labels to Regulate Risks

*Paul Weirich*

Many food products contain genetically modified (GM) corn and soybeans. Corn plants are modified to resist pests, and soybean plants are modified to tolerate herbicides. In the United States, food labels need not indicate GM ingredients. In the European Union, they must. Scientific data supports the safety of GM food, but consumers may want to exercise caution. What principles should guide a government's regulation of food labels? The popular Precautionary Principle is too single-minded to be a reliable source of sound judgment. In this chapter I propose a more thorough principle grounded in decision theory. It relies on cost–benefit analysis formulated comprehensively to assess a regulation's consequences. Made comprehensive, cost–benefit analysis absorbs plausible versions of the Precautionary Principle and wisely guides regulation.

## FOOD LABELS

Genetically modified food is widespread and on the increase for multiple reasons. Several new products may reach the marketplace soon. Genetically modified rice has beta-carotene, which stimulates production of vitamin A and so may reduce blindness in populations with diets reliant on rice. Genetically modified sorghum has additional amino acids that enhance nutritional value and may alleviate malnutrition in countries where that cereal is a staple. Coffee may be genetically modified to make beans ripen together so that mechanical harvesting is efficient. Soybeans may be genetically modified to remove an allergen that prevents some infants from drinking soy milk. Genetic modification may speed the growth of farmed salmon, may improve the quality of beef, and may make pigs more environmentally friendly.

Some farmers prefer raising GM crops because of economic benefits. Raising them reduces time in the fields spent tilling and spraying, reduces expenses for herbicides, pesticides, and fuel, and generates higher yields. The reduction in tilling preserves soil, and the reduction in use of herbicides, pesticides, and fuel also benefits the environment. Increases in food production help feed the globe's growing population.

Despite benefits, food derived from GM plants and animals prompts various adverse reactions. Some doubt the safety of GM food. Some fear the environmental consequences of GM crops. Genetic modification may spread through pollen drift. It may harm organisms feeding on plants affected by GM crops. Some oppose the corporations promoting GM food. They believe that those corporations have too much power. Some favor a lifestyle with reduced dependency on technology.

This chapter does not assess the cost and benefits of GM food.[1] It just considers whether GM food should be labeled. Labeling allows consumers to purchase according to their diverse attitudes toward GM food. I address the labeling of GM foods on the market rather than GM foods in the research pipeline. New GM foods, such as foods from GM animals, may create regulatory issues that do not arise for GM corn and soybeans.

In the United States, each state has some regulatory authority. A referendum in Oregon proposed labeling GM food, but voters defeated it. In Vermont, the state legislature considered several bills requiring that GM food be labeled, but none passed. Federal laws concerning interstate commerce make it difficult to sustain such state laws through inevitable court challenges. This chapter treats federal regulation of labeling in the United States, in particular, regulation by the Food and Drug Administration (FDA).

The FDA is the principal federal agency regulating food labels in the United States. The FDA issues guidelines for voluntary labeling of GM food but does not plan to require its labeling.[2] The FDA's policy is to require labeling GM food only if it is materially different from its conventional counterpart in nutrition, health, or use. Not finding scientific evidence of such differences, it does not require labeling. Under current labeling regulations, organic food may not contain GM ingredients. Some consumers purchase food labeled as organic to avoid GM food.

Assessing policies concerning food labels requires comparing them with alternative policies. Liability laws provide protection from food risks independently of regulatory law. Also, social opinion propagated in the media deters food companies from compromising food safety. The Internet supplements food labels as a source of information about food ingredients. Food labeling may be voluntary or mandatory. Mandates may require labeling to indicate that a product contains, may contain, or lacks GM ingredients. I consider chiefly mandatory labeling that says of products with GM ingredients that they contain those ingredients. Is mandatory labeling justified despite alternative responses to GM food?[3]

The case for mandatory labeling may proceed down various avenues. It requires justification of government imposition of labeling's cost. Consumers who do not want the information that labeling provides must still pay its cost in the form of increases in food prices. Some possible reasons for mandatory labeling are correction of market failure, consumer autonomy, and unknown risks. Political theory recognizes the legitimacy of government regulation for health and safety but debates regulation for social welfare and consumer autonomy. Hence, I explore the case for mandatory labeling to promote health and safety.

Labels allow consumers to avoid the health risks of GM food. Most research supports the safety of GM foods on the market, but new technologies are not risk-free. Genetic modification of crops may affect the composition of food. The effects of food's composition on health are not fully known. As Bucchini and Goldman (2002) and the National Research Council (2004) note, GM foods may cause allergic reactions. Genetically modified soybeans have nearly the same composition as conventional soybeans. However, StarLink corn is not approved for human consumption because it is not known whether the novel protein in it is an allergen. Premarketing tests screen for allergens, but errors are possible. The tests treat a relatively small sample. Postmarketing studies of consumption of a food product throughout the population over a long period may show that some approved products have adverse consequences for health.

If a food carries a risk, should it not be off the market rather than labeled? The FDA resorts to warning labels in special cases in which foods present a risk for a small segment of the population. Labels alert consumers that foods have or may have allergens from peanuts or wheat. Also, they signal the presence of transfat. A ban on transfat is impractical, and consumers may moderate its risks by reducing its consumption. Food that is risk-free is an impossible dream. The FDA bans foods with significant risks. How small a risk counts as negligible is a matter of judgment, however. Similarly, the reliability of premarketing scientific tests is a matter of judgment. Some reasonable consumers may be more or less cautious than the FDA. Labeling allows consumers to exercise choice about safety. It may be a sensible alternative to a ban.

The case against mandatory labeling in the United States cites its cost. Food labeling in the United States requires new methods of handling and storing grain. It requires separate systems for GM and non-GM crops. These new systems, which the agricultural corporation ADM is voluntarily creating, impose costs on food production. Also, labeling GM food may adversely affect the industry creating agbiotechnology. Labeling may lead consumers and retailers to reject GM foods enough that companies such as Monsanto reduce research on new GM crops. Inhibiting agbiotechnology withdraws its benefits from people around the globe. The European Union provides a case study of the effects of labeling requirements.[4] Few GM food products are on the market there.

Comparing the consequences of labeling and not labeling GM food requires assessing their consequences in the context of a nation's economy. Their assessment attends to farm subsidies and trade with other nations, using methods that, for example, Laffont and Tirole (1993) and Appleton (2000) present. The division between the United States and the European Union over labeling creates obstacles to trade. The European Union effectively bars GM food and, consequently, many U.S. food products. The United States holds that the European Union restricts trade in ways contrary to the policies of the World Trade Organization (WTO). The WTO requires the European Union to make payments to the United States because the European Union's ban on beef from cattle given recombinant bovine growth hormone does not rest on any scientifically established risk. It may do the same for GM food products not shown to carry any risk. Separating and labeling GM food will diminish these trade battles. Labeling may open markets to GM food from the United States.

Division over GM food entangles developing countries. They stand to benefit from crops genetically modified to enhance nutritional value, to increase resistance to drought and pests, and to reduce reliance on fertilizers and pesticides. Nonetheless, some developing countries share the European Union's aversion to GM food. For instance, Zambia refused a large donation of GM corn from the United States. It refused even given an offer to mill the corn to prevent its mixture with conventional corn seed.

Bernauer (2003: 61–63) reports a global trend toward greater restrictions on GM food and toward its mandatory labeling. Without labeling, resistance to GM food will mount. If Bernauer is right, labeling may be the best way to preserve the benefits of GM food. Labeling will allow GM food to establish its niche in the world.

## PRINCIPLES OF REGULATION

Sound principles for regulation take account of the context of regulatory decisions. This section sets the stage for such principles by reviewing the framework for federal regulatory decisions in the United States.

In a representative democracy, elected officials make decisions for the public. They use expert opinion to advance the public's interests. A principle of representative democracy instructs the legislature to enact laws that the public in ideal conditions would adopt if informed. Applying this principle to regulations for health and safety requires the legislature to use expert information about risks. The public, lacking information, may not agree with its representatives. For example, the government may declare a technology safe although the public harbors doubts. Representative democracy supplements public opinion with expert information.

Congress creates regulatory agencies such as the FDA and delegates to them decision-making responsibilities concerning public safety and health. A regulatory agency's charter issues mandates and imposes constraints. Typically, an agency's charter instructs it to use the best available information to impose regulations that protect the public's interests, provided that the regulations are economically feasible. Given that regulatory agencies should represent the public, they should make decisions that the public would make for itself in ideal conditions for public deliberation using expert information and decision-making tools. Which principles serve this goal?

Differences in U.S. and E.U. food policies arise because of differences in principles for regulating new technologies that may carry risks. The European Union favors the Precautionary Principle, and the United States favors cost–benefit analysis. Bernauer (2003: 167) suggests that, for food policy makers in the United States, "[t]he real target are *all* European environmental and consumer risk regulations based on the Precautionary Principle rather than on what U.S. regulators call 'sound science.'"

The following sections compare the Precautionary Principle and the principle of cost–benefit analysis. Which regulatory principle better articulates the principle of representative democracy according to which regulatory agencies are trustees for the public? Does representative democracy favor the Precautionary Principle or the principle of cost–benefit analysis? The next section points out problems with the Precautionary Principle. The section after it advocates a form of cost–benefit analysis that incorporates the virtues of the Precautionary Principle but also overcomes its limitations. Representative democracy supports that form of cost–benefit analysis.

## THE PRECAUTIONARY PRINCIPLE

The Precautionary Principle for regulatory decisions has the spotlight because of numerous endorsements by diverse organizations, such as the Commission of the European Communities (2000). Stated strongly, it prohibits any technology involving serious risks even if the likelihood of harm is unknown. Stated moderately, it notes that regulating a technology may be reasonable even in the absence of proof that the regulation will prevent harm. It urges regulatory agencies not to wait for proof of harm before taking steps to prevent it. A strong version of the Precautionary Principle bans GM food. A weak version of the principle licenses labeling so that consumers may take steps to avoid risks.

The Precautionary Principle embodies a good dose of common sense but nonetheless oversimplifies. Making decisions in the face of risk calls for a more subtle assessment of risks than it provides. Because proponents formulate the principle many ways, I consider two representative versions of the principle and point out the shortcomings of each.[5]

The Precautionary Principle appears in discussions of threats to the environment as well as threats to public safety and health. The first precise formulation I examine arises in a document on regulations to protect the environment, namely, the Rio Declaration on Environment and Development. It states the principle this way: "Where there are threats of serious or irreversible damage, lack of full scientific certainty shall not be used as a reason for postponing cost-effective measures to prevent environmental degradation" (United Nations, 1992: Annex I, Principle 15).

Cranor (1999) notices that this statement of the principle leaves ambiguous its exclusion of reasons against regulations. Does it deny that in regulatory matters the absence of evidence of damage is a relevant consideration, or only that it is a decisive reason? Also, the principle leaves unspecified the threshold of evidence envisaged. Legal practice suggests several possibilities. Should the threshold concern reasonable doubt or the preponderance of the evidence, for example? Cranor presents a more specific version of the principle that he regards as plausible: "Where there are threats of serious or irreversible damage, *failure to establish the threat of damage beyond a reasonable doubt* shall not be used as a *sufficient* reason for postponing cost-effective measures to prevent environmental degradation" (Cranor, 1999: 83, emphasis on "sufficient" added).

The Precautionary Principle is persuasive as Cranor states it. Nonetheless, it is an inadequate guide for regulation because it covers very little ground. It disqualifies certain reasons for not imposing regulations. However, it does not say when to impose regulations and when not to impose them. For example, it does not say whether any reasons together with the absence of convincing evidence of harm may suffice to block cost-effective regulation. Regulation needs a more decisive principle.

Sunstein (2003) concurs with Cranor about the plausibility of versions of the Precautionary Principle in the spirit of the Rio Declaration but looks for a stronger version of the principle with more substance. As a representative of various stronger versions of the principle in the literature, he considers this principle: "Regulation is required whenever there is a possible risk to health, safety, or the environment, even if the supporting evidence is speculative and even if the economic costs of regulation are high" (Sunstein 2003: 1018).

Sunstein moderates the principle by specifying that a possible risk in its sense requires scientific plausibility beyond a certain threshold. Nonetheless, this principle is much more demanding than is the Rio Declaration's principle. It requires extensive regulation. In the end, Sunstein (2003: 1020–1029) rejects the principle because its prohibitions are paralyzing. It is intolerant of risk and so in some decision problems forbids all responses, including inaction. In some cases, no course of action is risk-free and so acceptable according to the principle. For example, the principle may forbid introducing a new drug because testing is incomplete and may simultaneously forbid withholding the drug because people may suffer without it. According to Sunstein (2002: 136), the FDA should authorize drugs more readily to alleviate risks withholding them creates, in particular, the risk of denying a great health benefit. Risk is a pervasive feature of our world. Sound regulatory principles make concessions to its omnipresence.[6]

Senator Christopher ("Kit") Bond, a Republican from Missouri, puts the point this way: "The greatest risk associated with biotechnology is not to the monarch butterfly larvae, but from the naysayers, who may succeed in their goal to undermine biotech and condemn the world's population to unnecessary malnutrition, blindness, sickness, and environmental degradation."[7]

The Precautionary Principle as commonly formulated is either too weak to be a general guide to regulation, or else impossibly demanding. It urges caution, but many alternative principles also yield caution. What is the right way to be cautious? Reasonable people may differ in degree of caution. Within a representative democracy, caution is best served by a balanced response to risks that adjudicates differences in attitude to them. Cost–benefit analysis does this job better than the Precautionary Principle does.

## COST–BENEFIT ANALYSIS

Since the 1980s, an executive order has required a cost–benefit analysis for each regulation imposed by the federal government. Cost–benefit analysis explicitly attending to costs that are risks is sometimes called risk–benefit analysis. Cost–benefit analysis constitutes the most common approach to regulation in the United States, and in the European Union, it supplements applications of the Precautionary Principle. Cost–benefit analysis is sometimes criticized for ignoring important considerations but may be broadened to overcome such criticisms. I argue for a version of cost–benefit analysis that absorbs the wisdom in the Precautionary

Principle but sheds that principle's limitations. This section introduces a comprehensive form of cost–benefit analysis to guide regulation.[8]

The principle of cost–benefit analysis recommends the course of action that best balances costs and benefits. It advises bearing a risk if there are compensating benefits, or if alternatives run greater risks. In general, the analysis surveys options, evaluates the expected utility of each, and selects an option that maximizes expected utility. In *Decision Space* (Weirich 2001), I elaborate cost–benefit analysis for government regulatory decisions concerning public safety and health. This section explains the method and applies it to regulation of food labels.

A cost–benefit assessment of a proposed regulation considers the regulation's outcome given various possible states of the world. The expected-utility tree in figure 12.1 shows how the regulation's expected utility arises from consideration of its possible outcomes.

States 1 and 2 in figure 12.1 are mutually exclusive and jointly exhaustive and so form a partition. Each state has a probability, and each outcome has a utility. A state's probability weights the utility of the outcome it generates. The sum of the weighted utilities of the outcomes is the regulation's expected utility. Its expected utility is a means of comparing it with other regulations and with the absence of regulations.[9]

In this chapter, I assume that regulatory agencies should enact standards serving the public's interests. An agency should make decisions that the public would make for itself if it had the agency's expert information and used it rationally. A decision-making method with this objective requires an assessment of the public's goals. It needs the public's utility assignment to the possible outcomes of regulatory decisions. Assuming that each citizen has an equal voice, the public's utility assignment, according to utilitarian tradition, is a sum of the utility assignments of individuals.

The expected utility an agency assigns to a regulation should rest on expert information about the likelihood of its possible outcomes. The appropriate probability assignment to outcomes arises from pooling expert information and applying inductive logic, including statistical methods. Although citizens have conflicting beliefs, the relevant expert information washes out most conflicts and yields a common informed probability assignment. Building expert information into a cost–benefit assessment of a proposed regulation has two main steps. First,

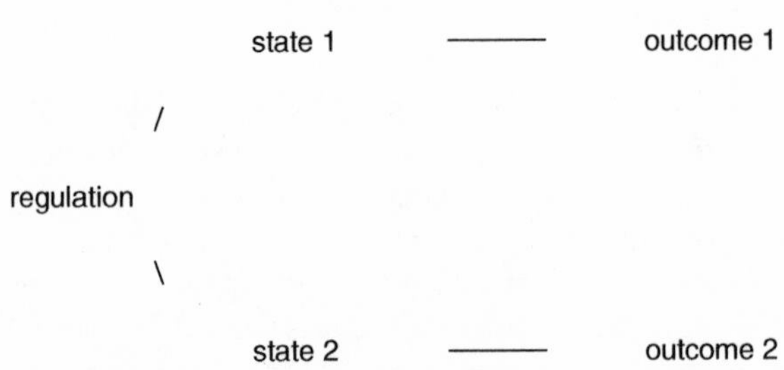


**Figure 12.1.** A regulation's expected utility.

to obtain the regulation's expected utility, it uses a fine-grained partition of states and assigns to each state in the partition an expert probability assignment. Second, it adopts a fine-grained specification of outcomes, such that further specification does not matter to any individual. The goal is an informed expected utility that is stable with respect to available information about relevant factors.

Cost–benefit analysis adds states to a regulation's expected-utility tree only to ensure that expert scientific information is brought to bear on evaluation of the regulation. It expands the tree only until utilities assigned to the outcomes at terminal nodes are stable with respect to additions of available information about objective, empirical facts. The probabilities of states are assessed using expert information. The utilities of outcomes are assessed from the public's perspective.[10]

It is crucial to take outcomes broadly. They must include everything that matters to people. Taking outcomes broadly makes them incorporate all factors to which responsive regulation attends. Some of those factors are information sensitive. Risks, for example, depend on probabilities, which are sensitive to information. People care about risks, and so outcomes include them. Because it takes outcomes broadly to include risk, a comprehensive version of cost–benefit analysis is sensitive to the public's aversion to risk (an aggregate of individual attitudes toward risk). That aversion lowers the expected utility of an option involving risk. The risk the option generates attaches to each of its possible outcomes and lowers their utilities. The larger the risk, the greater the decrement it generates.[11]

An outcome may realize a certain nutritional benefit together with a risk of allergic reaction. Then the outcome's utility is composite. A regulation's cost–benefit analysis involves not only an examination of the probabilities and utilities of its possible outcomes but also an investigation of the components of a possible outcome's utility. Using expert information to evaluate a proposed regulation requires adjusting and reevaluating the information-sensitive factors in an outcome. They must be reappraised to obtain the informed utility of the outcome. Calculating the outcome's informed utility requires breaking it down into its components. The appropriate analysis of the outcome's informed utility attends to the goals the outcome realizes.

The analysis of an outcome's utility in terms of the goals it realizes is traditional but not widely practiced. Two features are crucial. An outcome's utility is the sum of the *intrinsic* utilities of meeting the *basic* goals the outcome realizes. Basic goals are those from which other goals derive. Desires for basic goals need not be strong but are stable with respect to changes in information. Relying on them prevents double-counting considerations affecting an outcome's assessment. Intrinsic utility evaluates outcomes with respect to their logical consequences alone rather than their logical and causal consequences together. Its narrow scope also prevents double-counting considerations.

As a simple illustration, consider a regulatory agency deciding whether to approve two forms of GM food. Both offer the same probability of nutritional benefits and danger of allergic reaction. The main difference is that one has been more extensively tested. A larger body of statistical data supports the probability of its safety. A fully specified outcome's informed utility sums the intrinsic utility of its realization's consequences besides risk and the intrinsic utility of the risk it realizes. Knowing that the better tested food involves less risk and that the public has a basic aversion to risk, the informed utility assignments the agency calculates for that food's possible outcomes are higher than the informed utility assignments it calculates for the corresponding possible outcomes of the other food. Expected-utility calculations using informed utility assignments, calculated with intrinsic utility analysis, then prefer approval of the better tested food to approval of the less tested food.

A basic aversion to a risk is independent of information and is rational if proportionate to the risk. A nonbasic aversion to a risk depends on information and may be irrational because not appropriately responsive to available information. Assessing risks with respect to expert information corrects for mistaken nonbasic aversions to particular risks. It revises the public's uninformed assessments of risks. Expert information combined with the public's basic aversion to risk yields informed assessments of risks.

After introducing expert information, differences in attitude toward a GM food product may remain because of differences in basic attitudes toward risk. Politics, not science, must resolve those differences in attitude. A regulatory agency estimates the results of an ideal political process resolving those differences.

The way in which aversion to risk operates explains differences in attitudes toward different types of GM food. People are generally less tolerant of risks undertaken for the sake of flavor than risks undertaken for the sake of nutrition and health. Hence, they are more favorably disposed to genetic modification adding vitamins to rice than to genetic modification improving the flavor of tomatoes. Cost–benefit analysis evaluates each type of genetic modification according to attitudes to the particular the risk it imposes.

To illustrate the difference between the Precautionary Principle and the principle of cost–benefit analysis, consider the imposition of a regulation prohibiting genetic modification of rice to add vitamins. The strong version of the Precautionary Principle opposes the genetic modification because of the risk it inevitably creates. The principle supports the modification's prohibition even if science has not established that the modification will cause any harm. On the other hand, the principle of cost–benefit analysis rejects the regulation if, using the public's informed preferences, the likely benefits outweigh the risks.

The principle of cost–benefit analysis absorbs the weak version of the Precautionary Principle, which says not to prohibit cost-effective regulation protecting people against serious harm just because convincing evidence of harm has not been gathered. That version of the Precautionary Principle condones regulation without proof of harm, and cost–benefit analysis does the same. Even if science has not established that a genetic modification enhancing the flavor of tomatoes will harm people, the modification carries a risk. Given the public's aversion to risk, the absence of proof of harm is insufficient reason to halt regulation of the modification. According to the principle of cost–benefit analysis, blocking its regulation requires some great benefit that outweighs the public's aversion to the risk the modification creates. The principle of cost–benefit analysis provides for regulation without proof of harm and also specifies the circumstances that warrant that course of action.

A barrier to application of the principle of cost–benefit analysis in realistic cases is the absence of probability assignments to states and utility assignments to outcomes. Comprehensive cost–benefit analysis does not require using dollars to evaluate everything, including a human life. It uses only grounded value judgments and not ad hoc comparisons arising from a narrow view of costs and benefits. Following multiattribute analysis, it does not use a single scale to measure incomparable attributes. The method does not propose more precision than the grounds for probability and utility assignments warrant. Good (1952) generalizes expected-utility maximization to address the problem of imprecision. According to his generalization, a decision is rational if it maximizes expected utility under some *quantization* of beliefs and desires, that is, under some probability and utility assignments agreeing with beliefs and desires. This general principle uses nonquantitative beliefs and desires to constrain rational decisions as much as is warranted. Applying Good's principle, regulatory decisions may rest on nonquantitative probability and utility judgments. For example, although the probability that GM soybeans cause an allergic reaction is not known precisely, it may be known to be small enough to warrant approval of that GM food.

The FDA applies cost–benefit analysis comparatively in some cases. It does not regulate a GM food product if that product is as safe as another unregulated conventional food product. It assumes that the absence of a regulation for the conventional product is reasonable and therefore that the absence of a regulation for the GM product is reasonable, too. This comparative method does not require quantification of risks.[12]

Cost–benefit analysis, applied comprehensively, accommodates all considerations bearing on reasonable regulation of food labels, in particular, the public's aversion to risk. It steers a middle course between rashness and overcautiousness. It navigates between the Scylla and Charybdis confronting government regulation of risks and is a sensible guide for the regulation of food labels.

## AN AGENCY'S MANDATE

Comprehensive cost–benefit analysis makes a philosophical claim about reasonable regulation. It recommends a regulation offering the best balance of costs and benefits. It expresses a regulatory goal. In realistic cases impediments to attaining the goal provide excuses for falling short. For example, a regulation may fall short because of ignorance of a regulation's costs and benefits. A regulation may be justified despite falling short of the goal. Cost–benefit analysis's immediate practical consequences concern methods of justifying regulations. However, its application recommends revisions in regulations as well as their rationales.

The principle of cost–benefit analysis applies at various stages in the regulatory process. One application treats congressional mandates to regulatory agencies. The principle may guide congressional committees as, using the advice of experts, they draft the mandates of regulatory agencies. Another application treats a regulatory agency's execution of its mandate. The principle may guide regulatory agencies as they impose standards. A third application treats enforcement of standards.

The principle may guide field workers assessing compliance with standards. Each application raises issues the principle of cost–benefit analysis does not settle. Conflicting principles come to bear. To illustrate the principle, I simplify and put aside conflicting considerations such as human tendencies to err in its application. I show how the principle applies but do not claim that the principle's application is decisive. This section treats applications of the principle to the mandates of regulatory agencies. It treats issues that arise as a congressional committee drafts a regulatory agency's enabling legislation to specify the types of regulation the agency may impose and the reasons for which it may impose a regulation.

Two knotty issues arise immediately. The principle of cost–benefit analysis relies on a technical subject, expected-utility theory. In the United States, the regulatory agencies have a duty to publicly justify regulatory decisions. They must use nontechnical language suitable for the courts. How can their mandates incorporate the principle of cost–benefit analysis? Principles of regulation differ from principles for the public justification of regulations. The principle of cost–benefit analysis is a principle of regulation. It is only a guide to principles for the public justification of regulations. Principles for the justification of regulations, to gain simplicity, may be less precise than cost–benefit analysis. They may appeal to the consensus of scientists or to the judgment of a reasonable representative person. They may invoke the best available information instead of a probability assignment obtained by pooling expert information and applying inductive logic. Such simplifications are common and appropriate within legal practice. A regulatory agency's mandate may implicitly incorporate principles of regulation such as cost–benefit analysis by explicitly incorporating derivative principles for the public justification of regulation.

Also, although the principle of cost–benefit analysis expresses a goal of regulation, it may be best not to instruct a regulatory agency to pursue that goal directly. The agency's enabling legislation may be more effective in reaching the goal if it instructs the agency to pursue other goals that indirectly serve the main goal. For example, specifying pursuit of certain health goals and adoption of certain standards of evidence may be better than leaving to the agency discovery of the public's interests and the appropriate standards of evidence. Without specific directions, the agency may become mired in controversial foundational issues. The extent to which a regulatory agency's enabling legislation should explicate the general regulatory principles it serves is a topic for legal scholars.

This section uses cost–benefit analysis to address a dilemma about burden of proof that a mandate for a regulatory agency inevitably confronts. To present the dilemma, I review a controversy concerning the mandate of the Occupational Safety and Health Administration (OSHA). Mandates directing agencies to regulate food labels should learn from debates about other regulatory mandates. Past regulatory battles are cautionary tales for the regulation of food labels.

One regulatory goal is to impose a standard only if it brings benefits. Another regulatory goal is to impose a standard if it increases the likelihood of benefits. These goals conflict because often a regulation that promises benefits is not sure to yield them. For example, in 1978 OSHA imposed a new lower standard for exposure to benzene, a carcinogen. Its justification was the portion of its mandate that directed it to take all feasible steps to reduce risks to health. The American Petroleum Institute objected to the new standard. It pointed to the portion of OSHA's mandate that directed the agency to impose only standards that reduce impairment of health. Because science had not established that lowering exposure levels would prevent cancer, OSHA could not show that the new standard would reduce impairment of health, only that the standard was likely to reduce impairment of health. The two portions of OSHA's mandate were in conflict because of the difference between guaranteed and likely benefits.

The same factors bear on the FDA's regulation of food labeling. Should labels indicate whether a product contains ingredients from GM plants? The FDA's mandate gives it authority to require labeling of only material facts, which are defined as facts pertinent to nutritional profile or human health. Science has not established that consumption of GM corn and soybeans is material in this sense. It has not provided proof of harm. So the FDA does not require labeling for GM ingredients. Critics argue that it should require labeling because GM foods pose risks to human health, risks that exist as long as the complete safety of GM ingredients has not been conclusively established. Again, a conflict arises between regulating to avert harm and regulating to avert risks of harm.

Using a comprehensive cost–benefit analysis to resolve the conflict requires classifying risks. Risks fall into two main categories. First, there are physical risks that exist independently of the state of our knowledge. The probability on a coin toss of getting heads is a physical probability measured by the relative frequency of heads on coin tosses, and so the risk of losing a bet on heads is a physical risk. In contrast, the probability that a defendant on trial for a crime is guilty is an information-sensitive probability. It fluctuates as the defending and prosecuting attorneys take turns presenting evidence to the court. Hence the risk of convicting an innocent person is information sensitive.

The distinction between physical and information-sensitive risks rests on probability theory's distinction between physical and information-sensitive probabilities. Suppose that you receive a novel coin, and you assume that 50% is the probability of heads when it is tossed. Then you toss the coin many times and observe that heads turns up in 60% of the tosses. Because of the new information, you conclude that 60% is the probability of heads. Suppose that, indeed, the physical probability of heads is 60% because of the coin's composition and shape. This physical probability always was 60%. The change in your probability assignment from 50% to 60% was a change in the information-sensitive probability of heads. Corresponding to such physical and information-sensitive probabilities are physical and information-sensitive risks.[13]

A physical risk exists at a time. The risk at the time is independent of information. Given as much information as is physically possible at the time, the physical risk at the time remains. Full information reveals whether undertaking a physical risk of a harm results in the harm. However, full information about the risk's realization may be physically impossible at the time the risk is undertaken. At the time you bet on a coin toss, the physical risk of losing your bet is information independent. It remains given all the information it is physically possible to acquire at the time of the bet.

Ellsberg (1961) demonstrates that people have an aversion to information-sensitive risks. He offers the subjects in his studies choices between gambles in which they know the physical probabilities of outcomes and gambles in which they do not know the physical probabilities of outcomes. With the stakes held constant, a majority prefers gambling on events with known physical probabilities. The preference arises because of an aversion to information-sensitive risks, an aversion that increases with lack of information about underlying physical risks.

Genetically modified food generates both physical and information-sensitive risks. A study of the relative frequency with which a food product has a harmful side effect seeks to measure a physical risk. Uncertainty about the physical risk augments the information-sensitive risk. For instance, uncertainty about the physical risk of allergic reaction augments the information-sensitive risk of allergic reaction. An information-sensitive risk responds to the extent and credibility of testing, and thus may be lowered by engaging impartial scientists to ascertain the underlying physical risk.

Showing that a regulation reduces a physical risk of a harm requires showing that it reduces the relative frequency of the harm. Showing that a regulation reduces an information-sensitive risk of a harm requires showing only that the best available information does not preclude the regulation's reducing the physical risk of the harm. The public is averse to information-sensitive risks, and its attitude warrants regulation to reduce such risks according to a comprehensive version of cost–benefit analysis.

Reducing information-sensitive risks is not excessive caution. Reasonable people avoid such risks. For instance, reasonable drivers do not pass a truck on a hill whenever they fail to see oncoming traffic. They do not require proof of oncoming traffic before having a reason not to pass. When drivers refrain from passing on a hill, they respond to an information-sensitive risk and not knowledge of a physical risk. Drivers are ignorant of the physical risk. It is zero if there is no oncoming traffic. Although the physical risk of oncoming traffic is unknown, the information-sensitive risk of oncoming traffic justifies waiting for a clear view before passing. Similarly, regulatory agencies should not require grim proof that a new technology causes harm before taking steps to control it. It makes sense to reduce information-sensitive risks.

Congressional mandates concerning the regulation of food labels should be attuned to the difference between preventing a harm and reducing the risk of the harm. The difference may not seem significant because reducing the risk of a harm generally prevents the harm in the long run. Society is so vast that regulations apply to many cases. Reducing a risk of a harm thus prevents the harm in at least a few cases. For example, consider a regulation requiring labeling of GM corn. It reduces risk of allergic reaction. By reducing that risk, it prevents allergic reaction in some cases.

This step from risk reduction to harm prevention works only if the relevant risk is physical and concerns the relative frequency of some harm in a large population. It does not work if the relevant risk is information sensitive and arises from lack of information about the physical risk of a harm. Consuming GM corn is taking a chance not because one knows that harm results in a certain percentage of cases, but because one does not know that harm will not result. Reducing the information-sensitive risk does not ensure prevention of harm.

A congressional directive concerning the regulation of food labels should be formulated with court challenges in mind. If it says that regulations should prevent harm, then a regulation that reduces only an information-sensitive risk of a harm will be overturned in the courts. Proponents of the regulation will not have data demonstrating that reducing that risk will prevent the harm. Suppose, for instance, that Congress instructs the FDA to require food labels that reduce risk. What will be the burden of proof for a regulation responding to this directive? Must the FDA show that the regulation reduces a physical risk? Or will it be enough to show that the regulation reduces an information-sensitive risk of a harm created by a current lack of data? To be effective, the mandate must be clear about the burden of proof.

Comprehensive cost–benefit analysis recommends a mandate calling for standards that reduce informative-sensitive risks of impairment of health because the public is averse to such risks. A mandate responsive to the public's goals enables regulations that ward off information-sensitive risks. Rather than requiring proponents of a regulation to prove that it prevents a harm, a responsive mandate requires opponents of the regulation to prove that it does not prevent the harm despite reducing an information-sensitive risk of the harm. Advocates of the Precautionary Principle often have as a goal this shifting of the burden of proof to the case against regulation. The principle of cost–benefit analysis similarly achieves this shift provided that the principle recognizes information-sensitive risks, as it does when formulated comprehensively. It accommodates views about responsible risk management that influence proponents of the Precautionary Principle.

Regulation to reduce information-sensitive risks differs from regulation to reduce perceived risks. First, it uses assessments of risk that rest on the best available information. A perceived risk depends on a person's actual information. An information-sensitive risk may be assessed with respect to expert information. Cost–benefit analysis uses expert information to correct perceptions of risks. Second, an information-sensitive risk need not arise from perception of a corresponding physical risk. It may arise from uncertainty about the corresponding physical risk rather than from a belief about it. Cost–benefit analysis's regulation of information-sensitive risks is both more controlled and broader than regulation of perceived risks.

Applying comprehensive cost–benefit analysis to agency mandates requires assessing the public's goals. Congress has many methods of doing this. It uses polls, focus groups, agendas of public interest groups, lobbyists, the media, and the electoral process itself to learn what the public wants. Congress may express its assessment of the public's goals in its directives to regulatory agencies. For example, the common directive to take economically feasible steps to prevent impairment of health expresses Congress's belief that the public is willing to spend to protect health. The process of assessing the public's goals is imperfect. For example, vocal special-interest groups may gain more influence than they merit. Still, Congress's assessments form reasonable grounds for regulation.

Advances in cognitive psychology assist assessment of the public's attitudes toward risks. Kahneman and Tversky (1979) show that people are more willing to take risks to prevent harms than to gain benefits. Savage (1993) shows that the public's aversion to risk discriminates among types of risk. For example, an examination of the public's behavior shows that it is more averse to the risk of death from cancer than it is to the risk of death from highway accident. As Sunstein (2005: 140–53) notes, the public's aversion to a risk depends not only on the probability and seriousness of the possible harm but also on the type of possible harm. Applications of cost–benefit analysis should draw on experimental psychology to improve assessments of the public's attitudes toward risks. Similarly, it should draw on economics' studies of risk aversion. This discipline's literature on time preference, for instance, observes that aversion to risks that arise immediately is greater than aversion to risks of the same size that arise only in the remote future. Philosophy contributes by evaluating methods of aggregating individuals' attitudes to form the public's attitude and by assessing the rationality of attitudes toward risk. For example, it formulates principles about the consistency of attitudes toward risk and the proportionality of their strengths to the sizes of risks, as measured by the type, gravity, and probability of prospective harms.

Applying comprehensive cost–benefit analysis to food labeling illuminates the reasons for regulations. Some risks are physical, whereas other risks are information sensitive. Aversions to both types of risk are reasonable. Regulation legitimately targets both types of risk.

## BASELESS FEARS?

Representative democracy does not evaluate a rational citizen's basic goals. It takes them as sovereign. Basic goals persist despite changes in information. Hence, expert opinion does not trump them. Attitudes to risk are generally basic attitudes and are insensitive to information. Whether conduct is risky depends on information, but aversion to risky conduct does not depend on information. Aversion to information-sensitive risks is a basic attitude. Injection of expert information does not eliminate the aversion. Expert information changes the assessment of information-sensitive risks without eliminating aversion to them.

Should regulations reduce only physical risks involving physical probabilities, or may they also reduce information-sensitive risks involving information-sensitive probabilities? Will regulation of information-sensitive risks cater to baseless fears? Principles of representative democracy, such as principles of cost–benefit analysis, are not always decisive. However, other things being equal, a reduction in information-sensitive risk justifies a cost-effective regulation.

Early in the twentieth century, consumers feared food from crops modified through plant breeding that is now conventional. At that time, those food products did not present a physical risk, but they did present an information-sensitive risk. The fear was not baseless but rather a reasonable response to the information-sensitive risk. For all anyone knew then, the methods of plant breeding could have caused a change in the composition of food with ill effects for consumers. The information-sensitive probability of such a change was not insignificant. Fears may be rationally grounded in information-sensitive risks as well as in physical risks.

Whether GM food poses any physical risks to health will be settled by science in the long run. That GM food poses information-sensitive risks is clear now. Current science does not provide a proof of safety. Available information does not eliminate the epistemic possibility of harm. The information-sensitive probability of harm is not zero, and the corresponding information-sensitive risk is significant. Incomplete information about the physical risk of a harm creates an information-sensitive risk of the harm. The information-sensitive risk changes as research advances. Additional studies may reduce it. Statistics, the behavioral sciences, and decision theory offer objective methods of appraising the magnitude of information-sensitive risks with respect to the best available information. The courts may appeal to experts from these fields to establish the significance of information-sensitive risks that regulations target.

Given all available scientific information, GM food carries an information-sensitive risk because of the unknown long-term effects of consuming such food. If a consumer avoids GM food because of that risk, the consumer is reasonably responding to an objective, information-sensitive risk. A consumer without all available information, and so without an informed assessment of the magnitude of the information-sensitive risk of adverse health effects from GM food, may nonetheless reasonably believe that the risk is large enough to warrant avoiding GM food. For some consumers with reasonable attitudes toward risk, the information-sensitive risk, with respect to all available information, may not be negligible. If it is significant, it justifies mandatory labeling. Serving the public's interest requires reducing the risk to the extent feasible. Principles of representative democracy instruct an agency to reduce an information-sensitive risk that persists despite exposure to all available information, unless regulations are not cost-effective. Aversion to an information-sensitive risk is not uninformed or irrational. If feasible, a regulatory agency should facilitate removal of the object of the aversion.

Should the FDA's mandate be revised to target information-sensitive risks? Should it include new regulatory directives? Cost-effective standards that reduce risk are the goal of regulation. If Congress's mandate to the FDA (which does not explicitly incorporate cost–benefit analysis) prevents that agency's attainment of this goal, then, other things being equal, Congress should change the FDA's mandate. Of course, other things may not be equal. For the sake of gains elsewhere, say, in world trade, Congress may reasonably tolerate imperfections in the FDA's mandate. Putting aside the case for compromise, however, the FDA's mandate should direct it to reduce information-sensitive risks.

Does the FDA's mandate already target information-sensitive risks? The FDA's directives concerning food safety implicitly take account of information-sensitive risks. The FDA requires proof that a food is safe before condoning its general sale. In many cases, cost–benefit analysis endorses this caution. For example, it supports the FDA's halting the sale of beef after discovery of mad cow disease (bovine spongiform encephalopathy) in a herd. Even if the statistical data are too meager to reveal the physical risk of contracting variant Creutzfeldt-Jakob disease from consumption of beef from that herd, they establish a substantial information-sensitive risk to which the FDA appropriately responds.

However, in certain cases reasonable regulation requires a more nuanced assessment of information-sensitive risks than proof of safety entertains. Regulation needs direction from a cost–benefit analysis that explicitly recognizes information-sensitive risks and weighs them against benefits. If the potential benefits of a new GM food are great, they may outweigh information-sensitive risks. On the other hand, if the benefits are small, or unlikely to be realized, they may not warrant undertaking the information-sensitive risk its consumption involves. Comprehensive cost–benefit analysis suggests a revision of the FDA's regulatory standards to slow introduction of new foods when they carry information-sensitive risks. It explains which risks are reasonable.

Building cost–benefit analysis into the FDA's mandate increases its mandate's sensitivity to the reasons for and against regulations. Comprehensive cost–benefit analysis does not require elimination of information-sensitive risks but rather their reduction to a level at which potential rewards justify them. Cost–benefit analysis, when formulated to acknowledge information-sensitive risks, improves the FDA's regulation of food labels. It supports the FDA's regulation of some information-sensitive risks without making all such risks decisive. It balances risks and benefits and allows likely benefits to outweigh risks in select cases.[14]

Does the FDA's current enabling legislation give it authority to require mandatory labeling of GM ingredients? There are precedents for labels that indicate the process by which a food product was prepared rather than the food product's composition. Voluntary labeling advertises kosher foods, organic foods, dolphin-safe tuna, and diary products from cows not given the recombinant bovine growth hormone rBST. Mandatory labeling indicates irradiated meat and artificial flavors and colors. The FDA's labeling policy does not attend solely to the composition of a food product. In any case, because genetic modification may affect food composition and have unintended effects on health, the FDA's current mandate does not prohibit labeling requirements provided information-sensitive risks may serve as grounds for them. So cost–benefit analysis chiefly suggests revising its mandate to authorize more clearly regulation to reduce information-sensitive risks. A range of options for reducing those risks follows from that authorization.

## LABELING'S COSTS AND BENEFITS

Suppose that the FDA's mandate already includes or is revised to include cost–benefit analysis and the authority to require food labels to list GM ingredients because of information-sensitive risks. Does cost–benefit analysis recommend that labeling requirement? Are the costs of labeling GM food justified by the benefits?

Comprehensive cost–benefit analysis distinguishes physical and information-sensitive risks and gives each its due. Regulation does not require demonstration of physical risks. However, the pertinent science affects information-sensitive risks. Assessing information-sensitive risks with respect to the best available information diminishes their size. It weakens the case for regulations targeting such risks. If those risks do not justify taking GM food off the market, do they nonetheless justify mandatory labeling?

Risks may be reduced by prohibiting conduct that generates them, but also by furnishing information that allows consumers to avoid them. Labeling GM food allows consumers to reduce for themselves an information-sensitive risk. Consumers may use the information they have to evaluate information-sensitive risks that regulatory agencies identify using all available information. Leaving risk-reduction to consumers may be appropriate if available information does not reveal a physical risk that warrants removing a food product from the market, and reasonable attitudes to the information-sensitive risk vary widely, perhaps because of differences in aversion to risk. Mandatory labeling allows risk-averse people to avoid GM foods while permitting others to gain from their availability. By accommodating differences among consumers, it yields results better than either a ban on GM foods or the absence of regulations.

The most delicate step in applying cost–benefit analysis to labeling is identification of the public's informed goals. When GM food arrived a decade ago, approximately 300,000 people e-mailed the FDA asking that food labeled organic not be permitted to contain GM ingredients. The FDA was responsive and rescinded its decision to allow foods with GM ingredients to carry the organic label. The FDA may use such methods to guide its assessment of the public's informed goals.

Does the public desire to know whether food contains GM products? Do consumers want labeling of GM food? Survey data, although not conclusive, strongly suggest that the public wants labeling. The public's desire for labeling, however, may rest on ignorance of the safety of GM food. The public once aware of the material equivalence of conventional and GM soybeans may no longer care about knowing whether food contains GM soybeans, at least not sufficiently to bear the costs of labeling. Other information may also change the public's attitude toward GM foods. For example, Brown and Ping (2003) suggest that resistance to GM foods will decrease if consumers see personal benefits from these foods. In that case consumers will reduce their assessment of the risks and find them more acceptable. Also, the public may not trust the proof of safety the FDA has assembled for GM foods. The public may suspect that the experts consulted were biased. If so, then evidence of their impartiality may make the public less wary of GM food.

Identifying the public's informed goals requires careful attention to the public's basic goals, such as safety, and expert information about the best means of realizing those basic goals. A significant advantage of comprehensive cost–benefit analysis is its ability to combine the public's basic goals with expert information using the methods described above for cost–benefit analysis. Following those methods suggests that the information-sensitive risks of GM food are significant even when assessed with respect to the best available information. The FDA may reasonably mandate labeling GM foods until enough information about their safety reduces their information-sensitive risks so much that labeling's benefits no longer justify its costs. Although labeling's startup costs are high, the benefits in risk reduction may justify mandatory labeling with periodic reassessment of the case for it.

## MORALITY

The principle of cost–benefit analysis is a principle of rationality. It explains how to reach an objective of representative democracy, namely, regulations that the public would adopt in ideal conditions for negotiation given expert information. The principle of cost–benefit analysis is motivated by a hypothetical social contract but is not a principle of contractarian moral theory. It explicates the informed, rational will of the people, that is, their collective rationality. The public's hypothetical agreement may not express its current will because it is uninformed and not fully rational. The hypothetical conditions remove those obstacles to attaining goals of collective rationality. The policies of regulatory agencies should promote goals of collective rationality, which people would achieve if they were perfectly situated for joint action. This normative objective guides interpretation of the law and the law's application to particular cases.

Preceding sections of this chapter apply principles of representative democracy, articulated by comprehensive cost–benefit analysis, to the FDA's mandate and the FDA's regulation of food labels. A government agency should follow democratic principles unless other principles override them. Principles of morality occasionally override democratic principles. This section asks whether, all things considered, the FDA should require labeling GM food.

Moral principles bear on the issue of food labels in several ways. They explain what the law should be and also how current law applies to particular cases. Moral principles fill out the law. They direct interpretation of the law, as Dworkin (1986) explains. Applications of the law need the guidance of moral principles, especially in novel cases and in cases where the relevant parts of the law conflict.

The basic principle to act morally does not conflict with cost–benefit analysis because the public values morality. Morality is a benefit that comprehensive cost–benefit analysis acknowledges. However, the promotion of particular moral values, such as autonomy, may conflict with the recommendations of cost–benefit analysis. This section briefly reviews some moral considerations that an overall judgment about food labels must weigh.

After settling that a regulation accords with the informed, rational will of the people, a moral question remains: Is government authorized to impose that regulation? Citizens authorize democratic governments. Constitutions of democratic governments protect minorities against tyranny by the majority. Democracy does not authorize every law the public wants. A regulatory agency's duty to represent the public may conflict with its duty to respect an individual's rights against the majority.[15] Some courts have recognized a food corporation's First Amendment right to be silent about GM ingredients. Perhaps a consumer's right to exercise freedom of choice yields a countervailing right to informative labeling. Degnan (1997, 2000) and Streiffer and Rubel (2004) debate this issue. Because principles of democracy authorize regulation for health and safety, regulation of food labels has democracy's backing.

Morality considers the context of a regulatory law and its effects beyond consequences for health and safety. Cost–benefit analysis is reliably just only within just regulatory institutions. When regulatory institutions are imperfect, its results may need correction to compensate for the regulatory institutions' imperfections. Because a regulatory agency operates in an imperfect society, occasions arise in which it can alleviate problems outside its domain, for example, distributive injustices. Sometimes regulatory policy should compensate for social injustices, as Sunstein (2005: 50–51) notes. Labeling may reduce the market for GM food so that biotechnology slows and provides fewer benefits concerning nutrition, health, and the environment. Developing countries may suffer. Because genetic modification of food crops may prevent famine and blindness, consumers may have a moral obligation to tolerate the risks of GM food.

The extent of government's authority and its use of regulatory policy to compensate for social injustices are large issues. This chapter acknowledges their relevance to food labeling policy but does not address them. Its aim is just to apply the principles of representative democracy to labeling policy. It does not claim that those principles settle labeling policy.

## REGULATING INFORMATION-SENSITIVE RISKS

Genetically modified food offers great benefits but creates significant information-sensitive risks. These are risks whose size varies with the information available. They change in response to new data even if underlying physical risks are constant. Government regulatory agencies have the charge of protecting the public. Their mandates should instruct them to use expert information to serve the public's interests and, more specifically, to tolerate only information-sensitive risks justified by benefits. The Precautionary Principle is a sensible response to narrow cost–benefit analysis's neglect of information-sensitive risks. A better response, however, is a comprehensive form of cost–benefit analysis that acknowledges information-sensitive risks. Comprehensive cost–benefit analysis absorbs plausible versions of the Precautionary Principle. It yields a proportionate response to information-sensitive risks. Made comprehensive, cost–benefit analysis is a sound method of guiding government regulation of food labels.

## Notes

Thanks to Fred Degnan and Peter Markie for valuable comments on the manuscript.

1. For a survey of issues, see Beck (1995), McHughen (2000), Pence (2001, 2002), Nestle (2002), Schlosser (2002), Taylor and Tick (2003), Jasanoff (2005: Chap. 5), and Sunstein (2005: 27, 31, 64, 120, 128).
2. Brown and Ping (2003) discuss the FDA's policies for voluntary labeling of GM food. Use of the term "GM free" is discouraged because, for nearly all food products, it is technically inaccurate. Conventional plant breeding yields crops that are genetically modified, and traces of GM products occur adventitiously. To offset any insinuation that GM ingredients are unhealthful, the FDA requires a label advertising their absence to state that no scientific evidence shows that they are unhealthful.
3. Sunstein (2002: chap. 10) surveys social tools for controlling risks. Coleman (1992: part III) explains how tort law manages risks. Liability laws in Germany create a de facto ban on GM seed. Farmers are liable for any damage that results from its use.

4. For a review the European Union's labeling requirements, see Chege (1998), MacMaolain (2003), and Grossman (2005).
5. For additional discussion of the Precautionary Principle, see Miller and Conko (2001), Manson (2002), Posner (2004), Turner and Hartzell (2004), and Sunstein (2005).
6. Sunstein (2003: 1033–1032) mentions the possibility of substituting the Maximin Principle for the Precautionary Principle. The Maximin Principle recommends an act whose worst outcome is at least as good as the worst outcome of any alternative act. However, some objections to the Precautionary Principle carry over to the Maximin Principle. For example, as Binmore (1994: 317–327) argues, the Maximin Principle is not a reasonable response to risk. It is excessively cautious. Sunstein (2005: 109–115, 225) endorses an Anti-catastrophe Principle, which resembles the Maximin Principle but is more limited in scope. It protects against catastrophe even when the physical probability of catastrophe is unknown.
7. He spoke at the 2000 annual meeting of the American Association for the Advancement of Science in Washington, D.C., and was quoted in Environment News Service (2000).
8. Hansson (2005) is a critic of cost–benefit analysis. Sunstein (2005: 129), although not in favor of narrow cost–benefit analysis, argues that reasonable regulation attends carefully to costs and benefits. McGarity (1991: 11–13, chap. 9) contrasts cost–benefit analysis, which he calls regulatory analysis, with techno-bureaucratic rationality, which is procedural rather than substantive. He correctly observes that cost–benefit analysis involves many idealizations. However, as he concedes (164, chap. 20), it nonetheless offers useful guidance for regulatory decisions. Schmidtz (2002) is a defender of cost–benefit analysis.
9. Using obvious abbreviations, the regulation's expected utility is $P(S1)U(O1) + P(S2)U(O2)$. To illustrate, suppose that state 1 is three times as likely as state 2. In this case, the states' probabilities are, respectively, 3/4 and 1/4. Also, suppose that outcome 1 is twice as good as outcome 2, in which case the outcomes' utilities are, respectively, 2 and 1. Then the regulation's expected utility is $(3/4 \times 2) + (1/4 \times 1) = 7/4$.
10. A cost–benefit analysis applied to risks needs a probability assignment all citizens can endorse. It requires a single probability assignment that represents the rational opinion of all informed citizens concerning states of the world used to obtain a regulation's expected utility. Imagine that an expected-utility tree for a genetic modification represents as a possible state of the world that making the modification amounts to playing God. A religious group may hold the belief that the state obtains, whereas other groups hold the opposite belief. Appealing to expert information about the modification does not resolve the disagreement, so no common probability assignment to states emerges. To obtain a suitable probability assignment for states, cost–benefit analysis may exclude religious matters from the states used to calculate expected utilities and relegate religious matters to possible outcomes. Divergence about religious matters may then affect the utilities that individuals assign to outcomes without creating differences about probabilities of states.
11. I use the term "aversion to risk" in its ordinary sense, according to which it means aversion to the chance of a bad outcome. I do not give the term the technical sense it has in economics, where it signifies attributing to a commodity diminishing marginal utility.
12. Millstone et al. (1999) observe that the FDA's comparative method exempts GM food products from the rigorous testing that new drugs receive. Also, note that its method is not symmetric. It does not require labeling a new ingredient substantially equivalent to an ingredient whose labeling is required. For example, it does not require labeling a new natural flavor substantially equivalent to an artificial flavor that must be labeled.
13. For a comparison of physical and epistemic probabilities, see Hacking (2001: chap. 11). For an account of epistemic probabilities, the source of information-sensitive risks, see Skyrms (1986: sec. 1.6).

14. Regulatory reforms distinguishing physical and information-sensitive risks are not more difficult than other reforms that articles in law journals commonly recommend. For example, Wagner (1995) and Heinzerling (1998) suggest regulatory reforms aimed at separating scientific facts from policy and value judgments. They point out that arguments for regulations often misrepresent policy judgments as scientific facts. Heinzerling (1998: 1981, 2043–2056) observes that standard calculations of the costs of regulations per life saved involve controversial discounts of future benefits with respect to immediate benefits. Wagner (1995: 1628–1631) describes institutional pressures to resolve policy debates with appeals to science. She recommends that the courts require regulatory agencies to distinguish science from policy in their rationales for regulations (1711–1719).
15. Fear during World War II did not justify the U.S. internment of Japanese-Americans. The United States should have undergone the risk their freedom presented.

## References

Appleton, A. 2000. "The Labeling of GMO Products Pursuant to International Trade Rule." *NYU Environmental Law Journal* 8: 566–577.

Beck, U. 1995. *Ecological Politics in an Age of Risk*. Cambridge: Polity Press.

Bernauer, T. 2003. *Genes, Trade, and Regulation: The Seeds of Conflict in Food Biotechnology*. Princeton, N.J.: Princeton University Press.

Binmore, K. 1994. *Game Theory and the Social Contract: Vol. 1. Playing Fair*. Cambridge, Mass.: MIT Press.

Brown, J. L., and Y. Ping. 2003. "Consumer Perception of Risk Associated with Eating Genetically Engineered Soybeans Is Less in the Presence of a Perceived Consumer Benefit." *Journal of the American Dietetic Association* 103: 208–214.

Bucchini, L., and L. Goldman. 2002. "A Snapshot of Federal Research on Food Allergy: Implications for Genetically Modified Food." Pew Initiative on Food and Biotechnology. Available at http://pewagbiotech.org/research/allergy.pdf.

Chege, N. 1998. "Compulsory Labeling of Food Produced from Genetically Modified Soya Beans and Maize." *Columbia Journal of European Law* 4: 179–181.

Coleman, J. 1992. *Risks and Wrongs*. Cambridge: Cambridge University Press.

Commission of the European Communities. 2000. "Communication from the Commission on the Precautionary Principle." Available at europa.eu.int/comm/dgs/health_consumer/library/pub/pub07_en.pdf.

Cranor, C. 1999. "Asymmetric Information, the Precautionary Principle, and Burdens of Proof." In C. Raffensperger and J. Tickner, eds., *Protecting Public Health and the Environment: Implementing the Precautionary Principle*, pp. 74–99. Washington, D.C.: Island Press.

Degnan, F. 1997. "The Food Label and the Right to Know." *Food and Drug Law Journal* 52: 49–60.

Degnan, F. 2000. "Biotechnology and the Food Label: A Legal Perspective." *Food and Drug Law Journal* 55: 301–310.

Dworkin, R. 1986. *Law's Empire*. Cambridge, Mass.: Harvard University Press.

Ellsberg, D. 1961. "Risk, Ambiguity, and the Savage Axioms." *Quarterly Journal of Economics* 75: 643–669.

Environment News Service. February 23, 2000. "Franken Foods: Promise or Peril." Available at www.wired.com/print/science/discoveries/news/2000/02/34507.

Good, I. J. 1952. "Rational Decisions." *Journal of the Royal Statistical Society*, Series B, 14: 107–114.

Grossman, M. 2005. "Traceability and Labeling of GM Crops, Food, and Feed in the European Union." *Journal of Food Law and Policy* 1: 43–85.

Hacking, I. 2001. *An Introduction to Probability and Induction*. Cambridge: Cambridge University Press.

Hansson, S. June 21, 2005. "Social Decisions about Risk." Paper presented at the Conference on Philosophical Aspects of Social Choice, University of Caen.

Heinzerling, L. 1998. "Regulatory Costs of Mythic Proportions." *Yale Law Review* 107: 1981–2070.

Jasanoff, S. 2005. *Designs on Nature: Science and Democracy in Europe and the United States*. Princeton, N.J.: Princeton University Press.

Kahneman, D., and A. Tversky. 1979. "Prospect Theory." *Econometrica* 47: 263–291.

Laffont, J., and J. Tirole. 1993. *A Theory of Incentives in Procurement and Regulation*. Cambridge, Mass.: MIT Press.

MacMaolain, C. 2003. "The New Genetically Modified Food Labeling Requirements: Finally a Lasting Solution." *European Law Review* 28: 865–879.

Manson, N. A. 2002. "Formulating the Precautionary Principle." *Environmental Ethics* 24: 263–274.

McGarity, T. 1991. *Reinventing Rationality: The Role of Regulatory Analysis in the Federal Bureaucracy*. Cambridge: Cambridge University Press.

McHughen, A. 2000. *Pandora's Picnic Basket*. New York: Oxford University Press.

Miller, H. L., and G. Conko. 2001. "The Perils of Precaution: Why Regulators' 'Precautionary Principle' Is Doing More Harm Than Good." *Policy Review* 107: 25–39.

Millstone, E., E. Brunner, and S. Mayer. 1999. "Beyond Substantial Equivalence." *Nature* 401: 525–526.

National Research Council, Committee on Identifying and Assessing Unintended Effects of Genetically Engineered Foods on Human Health. 2004. *Safety of Genetically Engineered Foods: Approaches to Assessing Unintended Health Effects*. Washington, D.C.: National Academies Press. Available at books.nap.edu/catalog/10977.html.

Nestle, M. 2002. *Food Politics: How the Food Industry Influences Nutrition and Health*. Rockford, Ill.: Helm Publishing.

Pence, G. 2001. *Designer Food: Mutant Harvest or Breadbasket of the World?* Lanham, Md.: Rowman and Littlefield.

Pence, G., ed. 2002. *The Ethics of Food: A Reader for the Twenty-First Century*. Lanham, Md.: Rowman and Littlefield.

Posner, R. 2004. *Catastrophe: Risk and Response*. New York: Oxford University Press.

Savage, I. 1993. "An Empirical Investigation into the Effect of Psychological Perceptions on the Willingness to Pay to Reduce Risk." *Journal of Risk and Uncertainty* 6: 75–90.

Schlosser, E. 2002. *Fast Food Nation: The Dark Side of the All-American Meal*. New York: Harper Collins.

Schmidtz, D. 2002. "A Place for Cost-Benefit Analysis." In D. Schmidtz and E. Willott, eds., *Environmental Ethics: What Really Matters, What Really Works*, pp. 479–492. New York: Oxford University Press.

Skyrms, B. 1986. *Choice and Chance: An Introduction to Inductive Logic*, 3rd ed. Belmont, Calif.: Wadsworth.

Streiffer, R., and A. Rubel. 2004. "Democratic Principles and Mandatory Labeling of Genetically Engineered Foods." *Public Affairs Quarterly* 18: 223–248.

Sunstein, C. 2002. *Risk and Reason: Safety, Law, and the Environment*. Cambridge: Cambridge University Press.

Sunstein, C. 2003. "Beyond the Precautionary Principle." *University of Pennsylvania Law Review* 151: 1003–1058.

Sunstein, C. 2005. *Laws of Fear: Beyond the Precautionary Principle*. Cambridge: Cambridge University Press.

Taylor, M. R., and J. S. Tick. 2003. "Post-market Oversight of Biotech Foods: Is the System Prepared?" Pew Initiative on Food and Biotechnology. Available at pewagbiotech.org/research/postmarket/.

Turner, D. D., and L. C. Hartzell. 2004. "The Lack of Clarity in the Precautionary Principle." *Environmental Values* 13: 449–460.

United Nations. 1992. "Report of the United Nations Conference on Environment and Development." Rio de Janeiro, June 3–14, 1992. Available at www.un.org/documents/ga/conf151/aconf15126-1annex1.htm

Wagner, W. 1995. "The Science Charade in Toxic Risk Regulation." *Columbia Law Review* 95: 1613–1723.

Weirich, P. 2001. *Decision Space: Multidimensional Utility Analysis*. Cambridge: Cambridge University Press.

# Index